PE Civil: Water Resources and Environmental

Practice Exam with Solutions

10 warm-ups + 80 full exam

Copyright © 2024 by Yitong Pan. All rights reserved.

No part of this publication or any portion thereof may be reproduced, distributed, or transmitted in any form without the prior written permission of the author, except in the case of brief quotations embodied in certain noncommercial uses permitted by copyright law.

First Edition: May 2024

Printed in the United States of America

Preface

This PE Civil: Water Resources and Environmental Practice Exam with Solutions book contains a warm-up theoretical question set (10 questions) and a full sample 8-hour exam, including the morning Civil Engineering Breadth Test (40 questions) and the afternoon Water Resources and Environmental Depth Test (40 questions).

The question design of the book follows the NCEES Principles and Practice of Engineering Examination Civil: Water Resources and Environmental CBT Exam Specifications effective beginning in April 2024. The warm-up question set involves both theoretical and basic calculation questions. It is designed at an FE level so that readers can review the basic concepts that will be involved in the PE examination. The mock exam contains calculation questions only, intended to familiarize the readers with underlying concept understanding and reference manual usage. The mock questions aim to simulate the real PE examination difficulty level to provide a relatively accurate assessment before the exam.

The following reference resources will be available when accessing the examination:
- PE Civil Reference Handbook
- TSS Wastewater Facilities 2014
- TSS Water Works 2018

If you have any questions about the book, please feel free to reach out for an errata report or discussion at pecivil@outlook.com.

Good luck!

Warmup Questions + Solutions

10 Questions

1 hour

1. Sequencing the following wastewater treatment steps:

   (A) Primary clarifier
   (B) Disinfection
   (C) Secondary clarifier
   (D) Screening
   (E) Aeration / Activated sludge
   (F) Grit removal

Order	1	2	3	4	5	6
Steps						

2. Determine the coefficients of the following oxidation reaction that takes place in an industrial wastewater treatment process:

$$\_\_MnO_4^- + \_\_Fe^{2+} + \_\_H^+ \rightarrow \_\_Mn^{2+} + \_\_Fe^{3+} + \_\_H_2O$$

3. Which pollutant is typically associated with eutrophication in aquatic systems?

    (A) Nitrogen
    (B) Carbon
    (C) Phosphorus
    (D) Sulfur

4. Calculate the net excavation volume in cubic yards from the following data provided:

STATION	CROSS-SECTION AREA (ft²)	
	CUT	FILL
2+50	0	285
3+00	0	195
3+50	0	0
4+00	130	0

    (A) 625 yd³ fill
    (B) 505 yd³ fill
    (C) 350 yd³ fill
    (D) 120 yd³ cut

5. What is the moment of inertia to the center of a circle with a radius of 5 inches?

   (A) 79 in⁴
   (B) 491 in⁴
   (C) 982 in⁴
   (D) 1227 in⁴

6. Determine the force induced on string AO given that load C weighs 8 kg and angle θ = 37°.

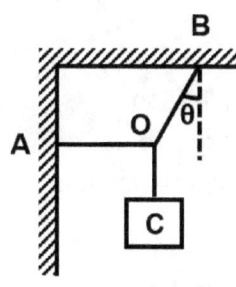

   (A) 60 N
   (B) 80 N
   (C) 100 N
   (D) 120 N

7. The following data was collected at a highway entrance. The average speed is most nearly:
   Hint: Round to one decimal place.

   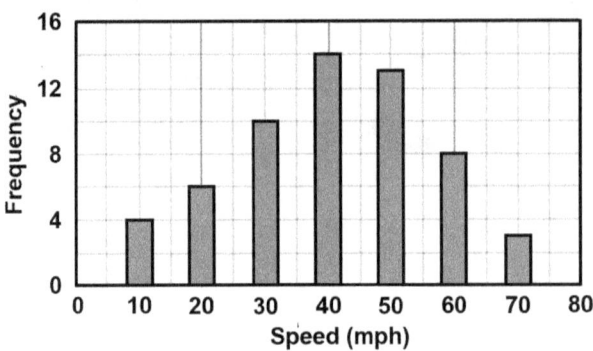

   Average speed: _____ mph

8. What is the purpose of a detention basin in stormwater management?

   (A) To increase infiltration rates
   (B) To remove pollutants through filtration
   (C) To provide habitat for aquatic organisms
   (D) To store excess stormwater and release it at a controlled rate

9. A rectangular channel section exhibits a width of 2 meters and a water depth of 1.4 meters. Determine the hydraulic radius of the channel.

   (A) 0.40 m
   (B) 0.52 m
   (C) 0.58 m
   (D) 0.82 m

10. A chemical reaction is determined to follow second-order kinetics with a rate constant of 0.3 L/mol·hr. If a reactant showed a resulting concentration of 1.58 mol/L after 20 minutes of reaction, determine the initial concentration of that reactant.

    (A) 1.88 mol/L
    (B) 1.75 mol/L
    (C) 1.68 mol/L
    (D) 1.62 mol/L

Solution 1

In wastewater treatment plants, the following steps are commonly involved in order:
1. Screening: remove large floating objects that may clog the equipment;
2. Grit removal: settle inorganic debris such as soil and small pebbles;
3. Primary clarifier: sink suspended solids to the bottom for sludge removal;
4. Aeration: degrade organic matter and remove nutrients from wastewater;
5. Secondary clarifier: separate treated wastewater from biological flocs;
6. Disinfection: eliminate remaining pathogenic microorganisms like bacteria and viruses;

The answer is:

Order	1	2	3	4	5	6
Steps	D	F	A	E	C	B

Solution 2

Determine the coefficients of the following oxidation reaction that takes place in an industrial wastewater treatment process:

$$\_\_MnO_4^- + \_\_Fe^{2+} + \_\_H^+ \rightarrow \_\_Mn^{2+} + \_\_Fe^{3+} + \_\_H_2O$$

The redox reaction can be divided into two half-reactions: oxidation and reduction
Oxidation: $Fe^{2+} - e^- \rightarrow Fe^{3+}$
Reduction: $MnO_4^- + 5e^- + 8H^+ \rightarrow Mn^{2+} + 4H_2O$
Since the electron gain and loss remain the same in the reaction, we have:
$$\begin{cases} 5Fe^{2+} - 5e^- \rightarrow 5Fe^{3+} \\ MnO_4^- + 5e^- + 8H^+ \rightarrow Mn^{2+} + 4H_2O \end{cases}$$
The final reaction: $MnO_4^- + 5Fe^{2+} + 8H^+ \rightarrow Mn^{2+} + 5Fe^{3+} + 4H_2O$

The answer is: **1, 5, 8, 1, 5, 4**

Solution 3

Which pollutant is typically associated with eutrophication in aquatic systems?

Eutrophication is a process in which water bodies become excessively enriched with nutrients, leading to an overabundance of algae and plants. This surge in plant growth depletes oxygen in the water, resulting in harmful algal blooms and detrimental effects on aquatic ecosystems.

Studies have consistently identified phosphorus and nitrogen as key contributors to eutrophication, with phosphorus being particularly significant due to its role as a limiting nutrient for algae growth and biomass accumulation. Consequently, effective management of phosphorus concentrations is essential for mitigating algae overgrowth and maintaining a stable environmental water quality. Therefore, phosphorous is typically recognized as the most important pollutant associated with aquatic eutrophication.

The answer is **(C)**

Solution 4

Calculate the net excavation volume in cubic yards from the following data provided:

STATION	CROSS-SECTION AREA (ft^2)	
	CUT	FILL
2+50	0	285
3+00	0	195
3+50	0	0
4+00	130	0

The cut or fill excavation volume (V) can be calculated through the formula: $V = L(A_1 + A_2)/2$
From the table, we can find that the distance between cross-sections L = 50ft; volume V:
$V_{fill} = \Sigma L(A_1 + A_2)/2 = 50ft \times (285ft^2 + 195ft^2) \div 2 + 50ft \times (195ft^2 + 0ft^2) \div 2 = 16875ft^3$
$V_{cut} = \Sigma L(A_1 + A_2)/2 = 50ft \times (0 + 130ft^2) \div 2 = 3250ft^3$
The fill volume (V_{fill}) is bigger than the cut volume (V_{cut}), therefore, the net excavation volume (V_{net}) should be dominated by the fill:
$V_{net} = V_{fill} - V_{cut} = 16875ft^3 - 3250ft^3 = 13625ft^3 = 504.6yd^3$

The answer is **(B)**

Solution 5

What is the moment of inertia to the center of a circle with a radius of 5 inches?

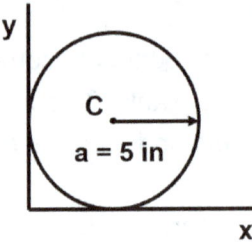

The moment of inertia (I_{x_C}) to the center of a circle can be calculated through the formula:
$$I_{x_C} = I_{y_C} = \pi a^4/4$$
Given a radius $a = 5in$, the moment of inertia I_{x_C}
$$I_{x_C} = \pi a^4/4 = 3.14 \times (5in)^4 \div 4 = 491 in^4$$

The answer is **(B)**

Solution 6

Determine the force induced on string AO given that load C weighs 8 kg and angle θ = 37°.

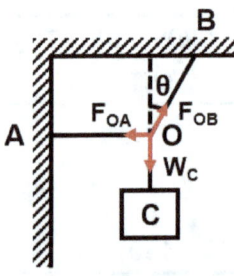

From the force analysis on Point O shown above, we can have the following relationships considering the sum of force on both axes (x-axis and y-axis) equals zero, $\sum F_{Ox} = \sum F_{Oy} = 0$:
$$\begin{cases} F_{OA} - F_{OB} \cdot \sin \theta = 0 \\ F_{OB} \cdot \cos \theta - W_C = 0 \end{cases}$$
load C gravity $W_C = mg = 8kg \times 9.8 N/kg = 78.4N$
Input all the information into the equation:
$$\begin{cases} F_{OA} - F_{OB} \cdot \sin 37° = 0 \\ F_{OB} \cdot \cos 37° - 78.4N = 0 \end{cases} \rightarrow \begin{cases} F_{OA} = 59.1N \\ F_{OB} = 98.2N \end{cases}$$
force induced on string AO $F_{AO} = F_{OA} = 59.1N$

The answer is **(A)**

Solution 7

The following data was collected at a highway entrance. The average speed is most nearly:

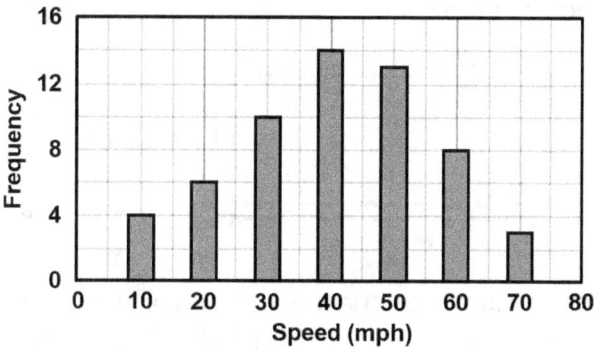

The average speed ($\bar{v}$) can be determined through the following equation:

$$\bar{v} = \frac{\Sigma(\text{speed} \times \text{frequency})}{\Sigma \text{frequency}}$$

$$\bar{v} = \frac{10 \times 4 + 20 \times 6 + 30 \times 10 + 40 \times 14 + 50 \times 13 + 60 \times 8 + 70 \times 3}{4 + 6 + 10 + 14 + 13 + 8 + 3} \text{mph} = 40.7 \text{mph}$$

The answer is **40.7**

Solution 8

What is the purpose of a detention basin in stormwater management?

A detention basin functions as a temporary reservoir for rainwater. During precipitation events, it retains the rainfall for a specified duration, rather than allowing it to immediately inundate the streets and strain the drainage infrastructure. This serves to mitigate flooding in urban areas.

Subsequently, upon cessation or attenuation of the precipitation, the water accumulated in the detention basin can be methodically discharged at a regulated pace. This deliberate discharge mechanism mitigates the risk of abrupt downstream inundation, which may otherwise engender flooding in adjacent locales or induce erosion along riverbanks.

The answer is **(D)**

Solution 9

A rectangular channel section exhibits a width of 2 meters and a water depth of 1.4 meters. Determine the hydraulic radius of the channel.

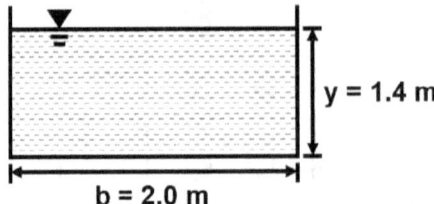

The hydraulic radius (R) for a rectangular channel can be calculated through the formula:

$$R = \frac{by}{b + 2y} = \frac{2.0m \times 1.4m}{2.0m + 2 \times 1.4m} = 0.583m$$

The answer is **(C)**

Solution 10

A chemical reaction is determined to follow second-order kinetics with a rate constant of 0.3 L/mol·hr. If a reactant showed a resulting concentration of 1.58 mol/L after 20 minutes of reaction, determine the initial concentration of that reactant.

For second-order reactions, we have the following relationship with regard to reaction time (t) and reactant concentrations (C_{A0}, C_A):

$$\frac{1}{C_A} - \frac{1}{C_{A0}} = kt$$

reaction time $t = 20\,min = 1/3\,hr$
Input all the information into the equation, we have:

$$\frac{1}{1.58M} - \frac{1}{C_{A0}} = 0.3 M^{-1} hr^{-1} \times (1/3\,hr) \quad \rightarrow \quad C_{A0} = 1.88M$$

The answer is **(A)**

Civil Breadth Questions

40 Questions

4 hours

1. Given the provided soil profile obtained during an onsite investigation, calculate the effective vertical stress at a depth of 24 feet.

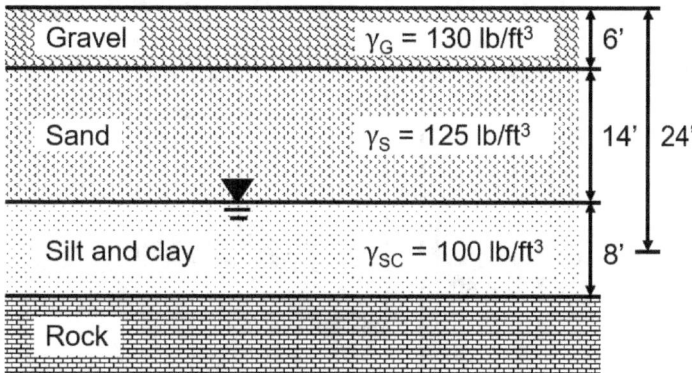

(A) 2680 psf
(B) 2780 psf
(C) 2930 psf
(D) 3180 psf

2. A saturated soil sample has a wet weight of 3.22 lb and a dry weight of 2.74 lb after drying in an oven. The specific gravity of the soil given from a practical test shows 2.4. Determine the void ratio of the soil sample.

(A) 2.40
(B) 0.36
(C) 2.82
(D) 0.42

3. Groundwater is passing through an inclined cylinder soil column at a flow rate of 3.4 ft³/min. The diameter of the column is 15 inches. Calculate the permeability of the soil according to Darcy's Law.

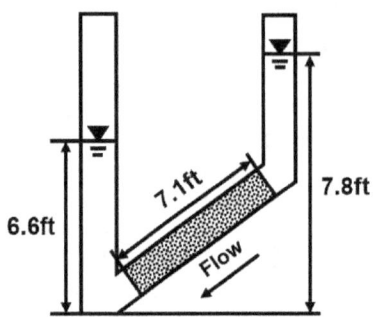

(A) 0.36 ft/s
(B) 0.68 ft/min
(C) 0.27 ft/s
(D) 0.40 ft/min

4. A truck is traveling on a highway at a speed of 66 ft/s when it experiences a sudden brake failure. The driver attempts to bring the truck to a stop by the nearby uphill with a grade of 0.2. Calculate the total distance the truck will travel before coming to a complete stop.

(A) 225 ft
(B) 344 ft
(C) 338 ft
(D) 425 ft

5. A sag vertical curve figure is shown below with the parameters provided. What are the lowest elevation of the curve and the elevation of PVT, respectively?

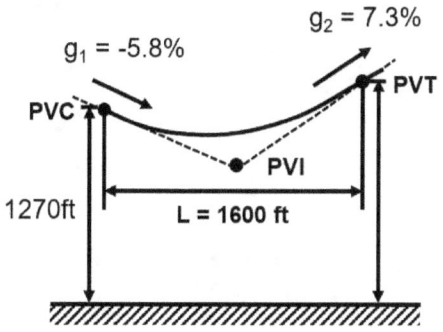

(A) 1249.8 ft; 1282 ft
(B) 1249.8 ft; 1294 ft
(C) 1249.5 ft; 1282 ft
(D) 1249.5 ft; 1294 ft

6. A 6-foot-long steel beam is subjected to a triangular distributed load shown below with w_{max} = 1.5 kips/ft. The beam has a rectangular cross-section with a width of 20 inches and a height of 30 inches. Determine the maximum deflection of the beam.

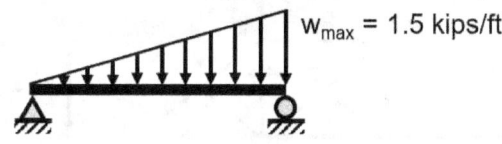

(A) 3.88×10^{-8} in
(B) 4.65×10^{-7} in
(C) 5.58×10^{-6} in
(D) 1.67×10^{-5} in

7. Assume an X-level flood has a return period of 35 years. Determine the possibility that a local 78-year-old has experienced flood event(s) at or above this magnitude.

   (A) 10.4%
   (B) 23.9%
   (C) 76.1%
   (D) 89.6%

8. Given a truss system supported over a span of 24 ft, identify the number of zero-force members within the system and calculate the axial force in member AB.

   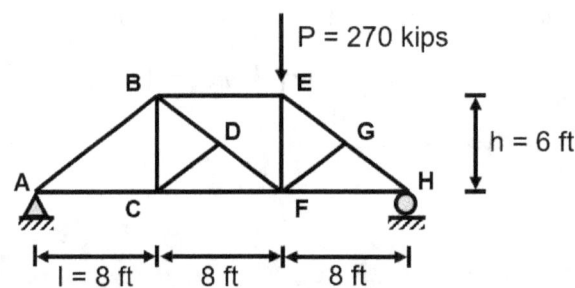

   (A) 3 zero-force members; 150 kips
   (B) 3 zero-force members; 225 kips
   (C) 2 zero-force members; 150 kips
   (D) 2 zero-force members; 225 kips

9. An engineer utilized 10 sacks of Portland cement to produce a grout mixture with a water-cement ratio of 0.8 ft³/sack. Assuming during the grout mixing, the cement volume shrank to half of its original volume while the water volume remained unchanged, determine the density of the applied cement.

   Hint: A sack of cement weighs 43 kg.

   (A) 53.8 pcf
   (B) 94.6 pcf
   (C) 118.3 pcf
   (D) 189.2 pcf

10. Water is discharged freely from a rounded orifice located 30 feet below the surface into the atmosphere. What is the velocity of the flow at the orifice opening?

    (A) 27 ft/s
    (B) 35 ft/s
    (C) 43 ft/s
    (D) 49 ft/s

11. Calculate the pressure at point A due to a square footing of width 0.5 ft and an applied load of 230 psf. Point A is located 12 inches below the base and 15 inches from the footing center.

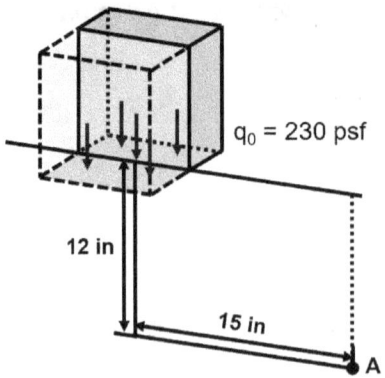

(A) 1.2 psf
(B) 2.3 psf
(C) 7.7 psf
(D) 11.5 psf

12. Analyze the traffic data gathered from a major highway during rush hour to calculate the peak hour factor.

TIME	NUMBER OF VEHICLES
00-07 min	1460
07-15 min	1595
15-22 min	2034
22-30 min	987
30-37 min	1639
37-45 min	2743
45-52 min	1956
52-60 min	1829

(A) 0.94
(B) 0.52
(C) 0.76
(D) 0.83

13. Bank soil was excavated from a specific borrow pit with a formation level of 32.5 ft before being trucked and compacted. The pit grid and spot heights (units in feet) obtained from a level survey are presented. Assume the soil has a swell factor of 0.2 and a shrinkage of 15%, calculate the final volume of the compacted soil.

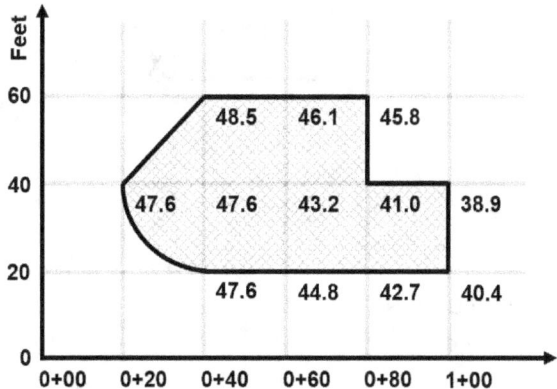

(A) 1048 yd³
(B) 1367 yd³
(C) 1139 yd³
(D) 1162 yd³

14. An investor is considering an engineering project with an expected future value of 2 million dollars in 25 years. If the annual interest rate is 4.8%, determine the initial investment (in dollars) the investor needs to make now to achieve this future value.

(A) 4.3×10^5
(B) 7.1×10^5
(C) 3.9×10^5
(D) 6.2×10^5

15. River water is passing through a trapezoidal open channel at a flow rate of 0.138 cfs. The cross-section of the channel is shown below. Assuming the water has a kinematic viscosity of 1.2 mm²/s, determine the type of the flow.

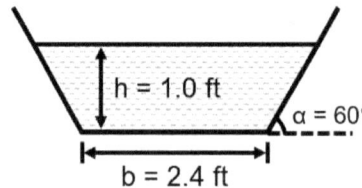

(A) Laminar flow
(B) Transitional flow
(C) Fully turbulent flow
(D) Lack of information provided

16. A highway soil material has 75 percent of the total passing through the 0.075 mm sieve. The group and plasticity index of the sample are 15.5 and 15 respectively. Determine the AASHTO classification of the material.

(A) A-2-7
(B) A-6
(C) A-7-5
(D) A-7-6

17. Following a storm, an average precipitation depth of 2.3 inches has been observed. The area has a curve number of 59, calculate the initial abstraction associated with the rainfall.

   (A) 1.4 inches
   (B) 2.2 inches
   (C) 3.1 inches
   (D) 6.9 inches

18. A concrete circular column with a length of 8 feet is fixed on both sides. Suppose the circular cross-section has a diameter of 20 inches and the column has a modulus of elasticity of 4,400 ksi. What is the critical stress of the column?

   (A) 236 ksi
   (B) 471 ksi
   (C) 942 ksi
   (D) 1885 ksi

19. Draw the shear diagram for the beam with the following loading conditions:

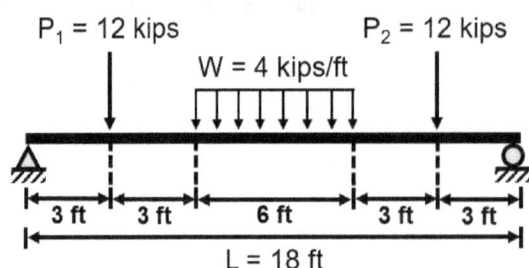

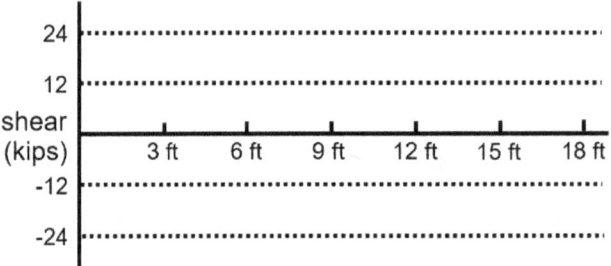

20. Determine the minimum time necessary to complete the entire industrial construction project based on the scheduled activity data.

ACTIVITY	PREDECESSOR	SUCCESSOR	DURATION IN DAYS
A	-	B,C	4
B	A	D	7
C	A	E	2
D	B,G	F	3
E	C	G	5
F	D	-	8
G	E	D,H	1
H	G	-	6

(A) 18 days
(B) 21 days
(C) 22 days
(D) 23 days

21. A retaining wall shown as follows is designed to support a cohesive backfill with a unit weight of 110 lb/ft³. Assuming an angle of internal friction of 30°, determine the active resultant force induced on the wall at a depth of 3.2 feet.

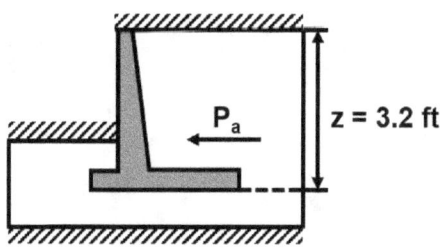

(A) 188 lb/ft
(B) 225 lb/ft
(C) 352 lb/ft
(D) 563 lb/ft

22. A wastewater treatment cylinder sedimentation tank has a diameter of 30 feet and a depth of 12 feet. Suppose the influent flow rate is 500,000 gallons per day, calculate the detention time in hours for the sedimentation tank.

(A) 6.08 hr
(B) 1.70 hr
(C) 3.04 hr
(D) 5.89 hr

23. A clay layer with a thickness of 8 feet is subjected to an effective stress of 1500 psf. Assume the clay has a compression index of 0.07 and an initial void ratio of 1.18. What is the consolidation settlement when an additional 1000 psf effective stress is loaded vertically to the top of the layer?

(A) 0.06 inches
(B) 0.26 inches
(C) 0.54 inches
(D) 0.68 inches

24. A 0.8-acre drainage pond is designed to collect rainwater from storm events with a retention time of 48 hours. Historical records indicate that the largest local storm events produced a rainfall intensity of 4.7 in/hr over an area of 3.3 acres. Assume the area has a runoff coefficient of 0.25. Determine the minimum depth required for the pond to hold the largest events.

(A) 4.7 ft
(B) 5.6 ft
(C) 19.4 ft
(D) 23.1 ft

25. A construction site requires dewatering to regulate the groundwater table with a 6-inch-diameter penetrating well. Detailed data regarding the unconfined aquifer is provided below. Assume the hydraulic conductivity estimated from the slug test is 36×10⁻⁴ cm/s. Determine the flow rate from the well given all the information provided.

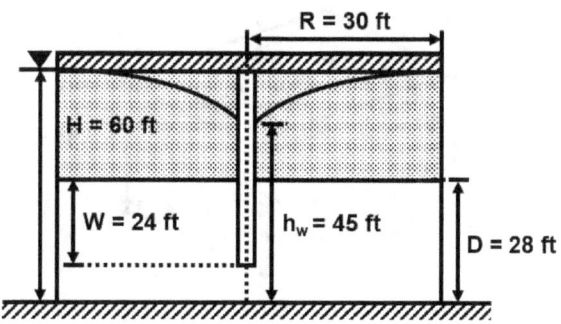

(A) 0.056 cfs
(B) 0.062 cfs
(C) 0.068 cfs
(D) 0.076 cfs

26. A geotechnical engineer is working on a project constructing a slope for the highway embankment. The slope is comprised of soil with a unit weight of 125 lb/ft³ and a cohesion of 50 kPa. Detailed information on the slope scale is provided below. Determine the factor of safety for this slope against sliding.

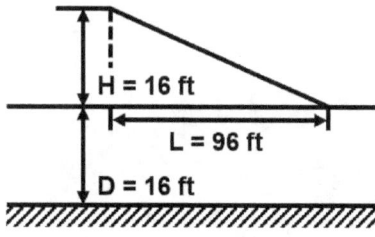

(A) 0.5
(B) 1.9
(C) 2.5
(D) 4.2

27. A 10" ×10" concrete column is subjected to an eccentric load of 64 kips including self-weight with an eccentricity of 2 inches to the right axis. Determine the maximum column compressive stress caused by this loading.

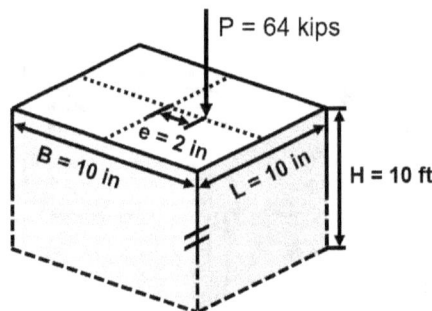

(A) 0.64 ksi
(B) 0.77 ksi
(C) 1.41 ksi
(D) 2.18 ksi

28. Given the same eccentric loading column shown in question 27, calculate the maximum bearing pressure on the column footing.

(A) 1.28 ksi
(B) 1.42 ksi
(C) 2.13 ksi
(D) 2.85 ksi

29. The water level in an aquifer is observed to decline from 129.6 inches to 129.3 inches over a distance of 370 feet. Suppose the aquifer has a specific discharge of 5.7 ft/day and a porosity of 0.4, what is the hydraulic conductivity and average seepage velocity of the aquifer?
Hint: Round to two decimal places.

Hydraulic conductivity: _____ ft/s
Seepage velocity: _____ ft/hr

30. A mobile crane with a self-weight of 110 kips and a telescoping boom weight of 10 kips is positioned on an industrial site with dimensions shown below. Given a total contact area of 25 ft², determine the average ground pressure when the crane reaches its maximum load capacity before tips over.

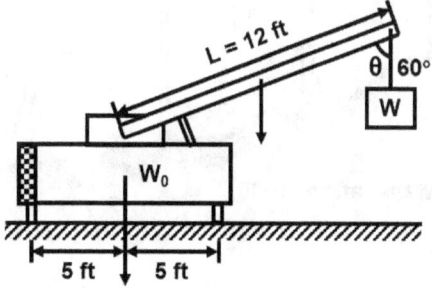

(A) 4.8 kips/ft²
(B) 6.4 kips/ft²
(C) 8.9 kips/ft²
(D) 9.3 kips/ft²

31. In a 1.6 ft radius circular pipe channel, water flows at a depth of 2.4 ft as shown in the figure. Given a Manning's roughness coefficient of 0.014 for the pipe material and a longitudinal slope of 0.25%, calculate the flow rate of water through the pipe.

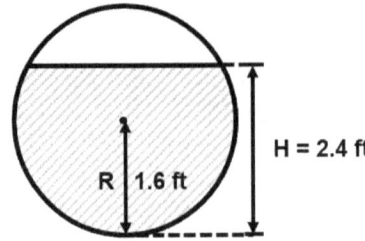

(A) 33.6 cfs
(B) 36.8 cfs
(C) 39.7 cfs
(D) 42.7 cfs

32. The survey leveling data gathered for a construction site is given below. Determine the ground elevation of point TP_2.

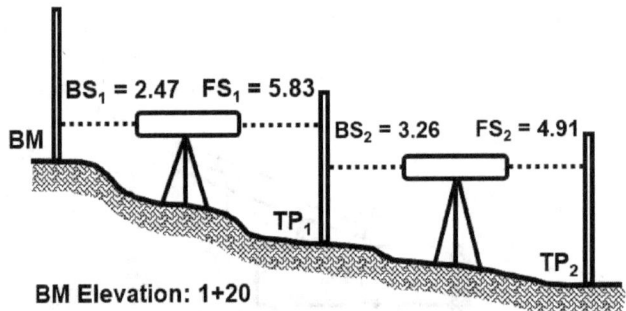

(A) 115.26 ft
(B) 114.17 ft
(C) 114.99 ft
(D) 113.48 ft

33. A city with a population of 2.1 million people is growing exponentially at a rate of 2.18%. According to the municipal water distribution on fire protection demands, determine the required fire flow in 8 years.

(A) 56.4 cfs
(B) 56.8 cfs
(C) 55.4 cfs
(D) 60.6 cfs

34. A table framing is shown below with supporting load information provided. Determine the total load on the 2-feet-long BII column.

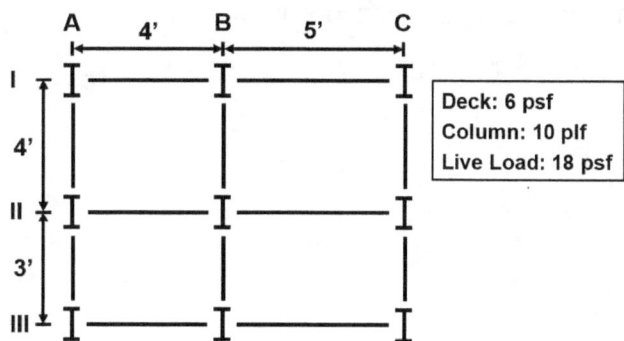

(A) 304 lb
(B) 344 lb
(C) 378 lb
(D) 398 lb

35. A fine sand layer and shallow wall foundation are shown below. With a combined weight of 85,000 lbf for the wall and the rigid circular footing, determine the surface vertical settlement. *Hint: Assume the sand layer has an elastic modulus of 180 tsf.*

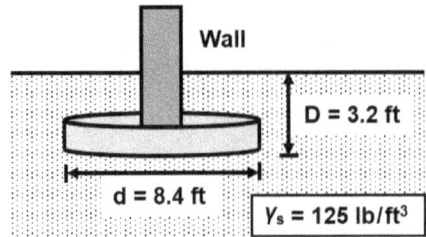

(A) 0.14 in
(B) 0.24 in
(C) 0.32 in
(D) 0.47 in

36. A water distribution system utilizes a 20-meter-long plastic circular pipe with a diameter of 0.6 m to convey water. The flow velocity through the pipe is 0.25 cm/s. Calculate the head loss due to friction in the pipe using the Hazen-Williams equation.

(A) 1.50×10^{-5} ft
(B) 1.15×10^{-6} ft
(C) 5.26×10^{-7} ft
(D) 4.04×10^{-8} ft

37. The house construction involves the following tasks shown with a table and a node notation. Suppose two groups of workers of 4 (workers not transferrable) are undertaking the entire construction project, and the cost per day per worker is $200. Estimate the minimum duration and the cost of the project.

    Hint: The duration in the table per task is the time required for two groups working together.

ORDER	TASK	DURATION (DAYS)
A	Site Preparation	10
B	Foundation laying	15
C	Construction	30
D	Interior finishing	20
E	Landscaping	10

    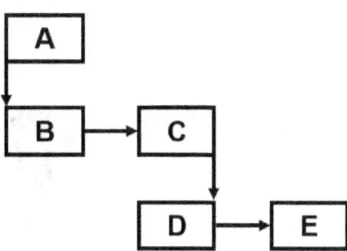

    (A) 55 days, $88000
    (B) 85 days, $88000
    (C) 55 days, $136000
    (D) 85 days, $136000

38. A water remediation project employs a specific adsorbent to reduce PFAS contamination in an open stream, achieving an adsorption efficiency of 80%. If the initial PFAS concentration is 10 mg/L and the water is treated at a flow rate of 500 gallons per day, calculate the total mass of PFAS removed over 15 days.

    (A) 0.13 lb
    (B) 0.34 lb
    (C) 0.50 lb
    (D) 0.62 lb

39. 8 solid pipes shown below are required to be installed for a new construction site. Suppose the inner diameter of the steel pipe is 10", and the costs of the steel and the filled concrete are 28 LF and 110 CY. Determine the total cost of the pipes needed.

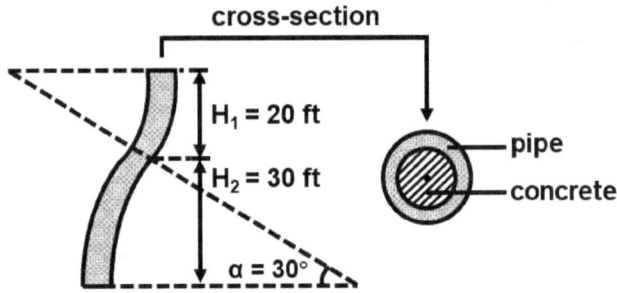

(A) 14113
(B) 12088
(C) 12660
(D) 13182

40. Certain components of a municipal water system are shown below. Tap water passed through a fully opened gate valve into a narrower tube, navigated two short-radius elbows, a pressure regulating valve, and exited the tubing. Given an inlet flow rate of 5 GPM and a total minor head loss of 0.2 meters, determine the pressure regulating valve head loss coefficient.

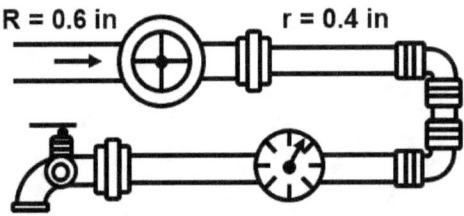

(A) 1.31
(B) 2.50
(C) 3.58
(D) 1.17

Civil Breadth Solutions

Answer Keys

Question	Answer	Question	Answer
1.	A	21.	A
2.	D	22.	C
3.	C	23.	D
4.	B	24.	C
5.	C	25.	B
6.	D	26.	D
7.	D	27.	C
8.	A	28.	B
9.	B	29.	0.98; 0.59
10.	C	30.	C
11.	B	31.	A
12.	C	32.	C
13.	D	33.	B
14.	D	34.	D
15.	C	35.	B
16.	C	36.	B
17.	A	37.	D
18.	B	38.	C
19.	N/A	39.	C
20.	D	40.	A

Solution 1
Given the provided soil profile obtained during an onsite investigation, calculate the effective vertical stress at a depth of 24 feet.

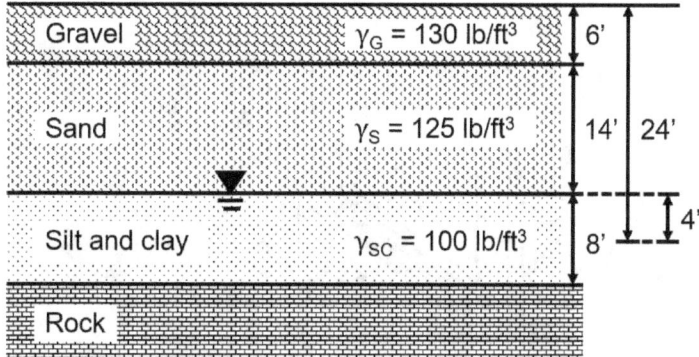

The effective vertical stress (P_{eff}) can be calculated by summing up the accumulated gravity stress minus the buoyant force caused by the presence of the water table: $P_{eff} = \sum \gamma H - \gamma_w h$
At the depth $d = 24ft$:
effective depth of the silt and clay layer $H_{SC} = d - H_G - H_S = 24ft - 6ft - 14ft = 4ft$
effective vertical stress $P_{eff} = \sum \gamma H - \gamma_w h = \gamma_G \times H_G + \gamma_S \times H_S + \gamma_{SC} \times H_{SC} - \gamma_w \times H_{SC}$
$P_{eff} = (130 \times 6 + 125 \times 14 + 100 \times 4 - 62.4 \times 4) lb/ft^2 = 2680.4 psf$

The answer is **(A)**

Solution 2
A saturated soil sample has a wet weight of 3.22 lb and a dry weight of 2.74 lb after drying in an oven. The specific gravity of the soil given from a practical test shows 2.4. Determine the void ratio of the soil sample.

The void ratio (e) of the soil sample can be determined through the formula:
$$e = \frac{W_w G}{W_s S}$$
weight of water $W_w = W_t - W_s = 3.22lb - 2.74lb = 0.48lb$
Since the sample is saturated, the degree of saturation $S = 1.00$
void ratio e
$$e = \frac{W_w G}{W_s S} = \frac{0.48lb \times 2.4}{2.74lb \times 1.00} = 0.42$$

The answer is **(D)**

Solution 3

Groundwater is passing through an inclined cylinder soil column at a flow rate of 3.4 ft³/min. The diameter of the column is 15 inches. Calculate the permeability of the soil according to Darcy's Law.

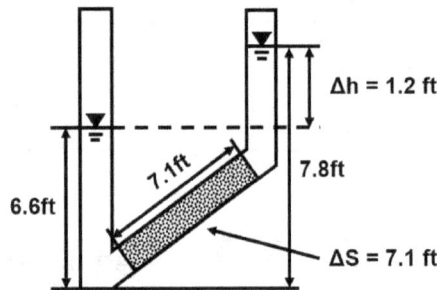

Darcy's Law represents the relationship between flow rate (Q), permeability (k), cross-sectional area (A), and hydraulic gradient (i), expressed by the formula:

$$Q = kiA \quad \rightarrow \quad k = Q/iA$$

hydraulic gradient $i = \Delta h/\Delta S = (7.8ft - 6.6ft) \div 7.1ft = 0.169$
cross-sectional area $A = \pi D^2 / 4 = 3.14 \times (15in)^2 \div 4 = 176.6 in^2 = 1.23 ft^2$
permeability $k = Q/iA = 3.4 ft^3/min \div (0.169 \times 1.23 ft^2) = 16.4 ft/min = 0.273 ft/s$

The answer is **(C)**

Solution 4

A truck is traveling on a highway at a speed of 66 ft/s when it experiences a sudden brake failure. The driver attempts to bring the truck to a stop by the nearby uphill with a grade of 0.2. Calculate the total distance the truck will travel before coming to a complete stop.

The total distance (d) the truck traveled before a complete stop can be calculated by determining the horizontal travel distance (d_b) first:

$$d_b = \frac{v_1^2 - v_2^2}{30(f + G)}$$

initial speed $v_1 = 66 ft/s = 66 ft/s \times 3600 s/h \div (5280 ft/mi) = 45 mph$; final speed $v_2 = 0 mph$
Since the brake failed to work, the deceleration of vehicle $a = 0$, leads to the coefficient of friction $f = a/g = 0$; the grade $G = 0.2$
horizontal travel distance d_b

$$d_b = \frac{v_1^2 - v_2^2}{30(f + G)} = \frac{45 mph^2 - 0 mph^2}{30 \times (0 + 0.2)} = 337.5 ft$$

total travel distance d

$$d = d_b\sqrt{1 + G^2} = 337.5 ft \times \sqrt{1 + 0.2^2} = 344.2 ft$$

The answer is **(B)**

Solution 5

A sag vertical curve figure is shown below with the parameters provided. What are the lowest elevation of the curve and the elevation of PVT, respectively?

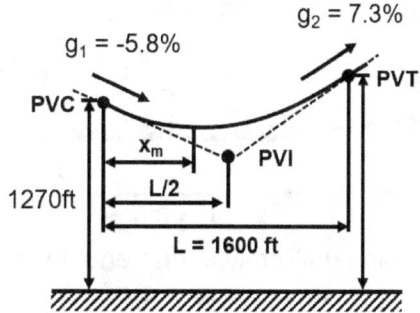

The lowest elevation of the curve (Y_m) could be determined by finding out the horizontal distance from PVC to the minimum elevation on the curve (X_m) first and calculating the curve elevation.

horizontal distance to min elevation on the curve X_m

$$X_m = \frac{g_1 L}{g_1 - g_2} = \frac{-5.8\% \times 1600 ft}{-5.8\% - 7.3\%} = 708.4 ft$$

curve elevation at the minimum point Y_m

$$Y_m = Y_{PVC} + g_1 X_m + \frac{X_m^2 (g_2 - g_1)}{2L}$$

$$Y_m = 1270 ft + 5.8\% \times 708.4 ft + (708.4 ft)^2 \times [7.3\% - (-5.8\%)] \div (2 \times 1600 ft) = 1249.5 ft$$

Please note that the PVI corresponding point on the curve is not the lowest when the absolute values of g_1 and g_2 are not equal.

The elevation of the PVT (Y_{PVT}) can be determined by the elevation of the PVC (Y_{PVC}). As a result of the equal tangent quality of the vertical curve, PVC and PVT are equidistant from the PVI:

$$Y_{PVT} = Y_{PVC} + (g_1 + g_2)L/2 = 1270 ft + (-5.8\% + 7.3\%) \times 1600 ft \div 2 = 1282 ft$$

The answer is **(C)**

Solution 6

A 6-foot-long steel beam is subjected to a triangular distributed load shown below with w_{max} = 1.5 kips/ft. The beam has a rectangular cross-section with a width of 20 inches and a height of 30 inches. Determine the maximum deflection of the beam.

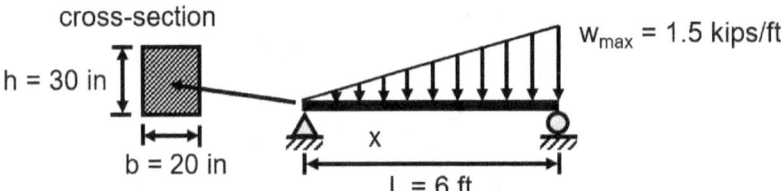

Considering the beam load increasing uniformly to one end, the maximum deflection (Δ_{max}) of the beam can be calculated through $\Delta_{max} = 0.0130WL^3/EI$
total load on beam $W = w_{max}L/2 = 1.5\text{kips/ft} \times 6\text{ft} \div 2 = 4.5\text{kips}$
length $L = 6\text{ft} = 6\text{ft} \times (12\text{in/ft}) = 72\text{in}$
Since the beam is made of steel, the modulus of elasticity of steel $E = 29000\text{ksi}$
moment of inertia to the beam center $I = bh^3/12 = 20\text{in} \times (30\text{in})^3 \div 12 = 45000\text{in}^4$
maximum deflection Δ_{max}

$$\Delta_{max} = \frac{0.0130WL^3}{EI} = \frac{0.0130 \times 4.5\text{kips} \times (72\text{in})^3}{29000\text{ksi} \times 45000\text{in}^4} = 1.67 \times 10^{-5}\text{in}$$

The answer is **(D)**

Solution 7

Assume an X-level flood has a return period of 35 years. Determine the possibility that a local 78-year-old has experienced flood event(s) at or above this magnitude.

Given the flood has a return period of $T = 35\text{yrs}$, the probability of a single occurrence in a given storm period $p = 1/T = 1 \div 35 = 0.0286$
Since the given period $n = 78$ years, the probability of the 78-year-old experiencing an exceeding flow $P = 1 - (1-p)^n = 1 - (1-0.0286)^{78} = 89.6\%$

The answer is **(D)**

Solution 8

Given a truss system supported over a span of 24 ft, identify the number of zero-force members within the system and calculate the axial force in member AB.

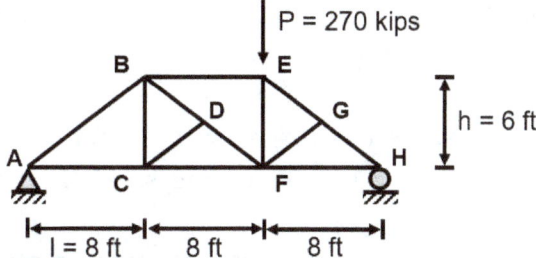

The zero-force members of the system can be determined through the following two rules:

I. If two non-collinear members meet at an unloaded joint, then both of them are zero-force;
II. If three forces (interaction, reaction, or applied forces) meet at a joint, and two of them are collinear, then the third member is a zero-force member.

Therefore, the zero-force members in the truss are CD (BF, rule II), FG (EH, rule II), and BC (AF, rule II), 3 members in total.

The truss system can then be simplified for calculating the axial force in member AB:

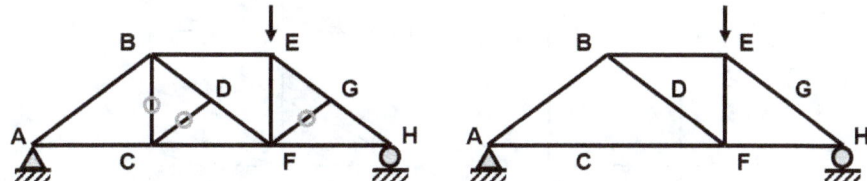

For point H, the sum of moments is 0; for point A, the sum of forces (y-axis) is 0:

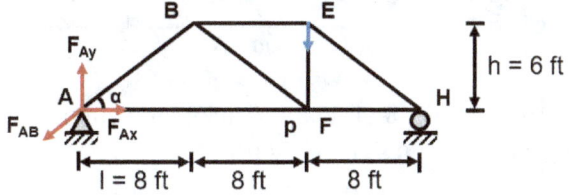

Point H: $\sum M_H = -Pl + 3F_{Ay}l = -270\text{kips} \times 8\text{ft} + 3F_{Ay} \times 8\text{ft} = 0 \quad \rightarrow \quad F_{Ay} = 90\text{kips}$

The opposite force of F_{Ay} can only be provided by member AB on the y-axis, according to the given information, $\tan \alpha = 6/8$, $\alpha = 37°$, $\sin \alpha = 3/5 = 0.6$

Point A (y-axis): $\sum F_A = F_{Ay} - F_{AB}\sin\alpha = 90\text{kips} - F_{AB} \times 0.6 = 0 \quad \rightarrow \quad F_{AB} = 150\text{kips}$

The answer is **(A)**

Solution 9

An engineer utilized 10 sacks of Portland cement to produce a grout mixture with a water-cement ratio of 0.8 ft³/sack. Assuming during the grout mixing, the cement volume shrank to half of its original volume while the water volume remained unchanged, determine the density of the applied cement.

The density of the cement (ρ_c) can be determined with the formula $\rho_c = m_c/V_c$; based on the known information, cement weight $m_c = 10\,\text{sacks} \times 43\,\text{kg/sack} \times 2.2\,\text{lb/kg} = 946\,\text{lb}$

To determine the cement volume (V_c), the provided information states that "the cement volume shrank to half of its original volume while the water volume remained unchanged during the grout mixing", so we have grout volume $V_g = 0.5V_c + V_w \quad \rightarrow \quad V_c = 2(V_g - V_w)$

The grout volume (V_g) can be determined through the "Cement Content of Portland-Cement Grout Mixes" graph provided in the reference manual: At a water-cement ratio (w/c) of 0.8 ft³/sack, the grout volume $V_g = 13\,\text{ft}^3$ when we have 10 sacks of cement;

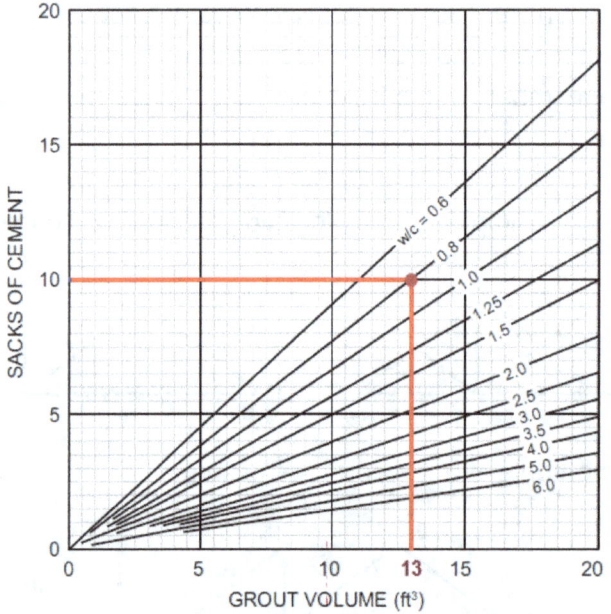

According to the formula: w/c = cubic feet water/sacks of cement
water volume $V_w = (w/c) \times 10\,\text{ft}^3 = 0.8 \times 10\,\text{ft}^3 = 8\,\text{ft}^3$
cement volume $V_c = 2(V_g - V_w) = 2 \times (13\,\text{ft}^3 - 8\,\text{ft}^3) = 10\,\text{ft}^3$
cement density $\rho_c = m_c/V_c = 946\,\text{lb} \div 10\,\text{ft}^3 = 94.6\,\text{pcf}$

The answer is **(B)**

Solution 10

Water is discharged freely from a rounded orifice located 30 feet below the surface into the atmosphere. What is the velocity of the flow at the orifice opening?

The flow rate (Q) at the orifice opening can be determined through the formula:

$$Q = C_d A_0 \sqrt{2gh}$$

Since the flow velocity $v = Q/A_0$

$$v = C_d \sqrt{2gh}$$

From the Orifices Nominal Coefficients charts in the reference manual, the coefficient of discharge (C_d) for a rounded orifice is 0.98

	SHARP EDGED	ROUNDED	SHORT TUBE	BORDA
C_d	0.61	0.98	0.80	0.51
C_c	0.62	1.00	1.00	0.52
C_v	0.98	0.98	0.80	0.98

Orifices and Their Nominal Coefficients

flow velocity v:

$$v = C_d \sqrt{2gh} = 0.98 \times \sqrt{2 \times 32.2 \text{ft/s}^2 \times 30 \text{ft}} = 43 \text{ft/s}$$

The answer is **(C)**

Solution 11

Calculate the pressure at point A due to a square footing of width 0.5 ft and an applied load of 230 psf. Point A is located 12 inches below the base and 15 inches from the footing center.

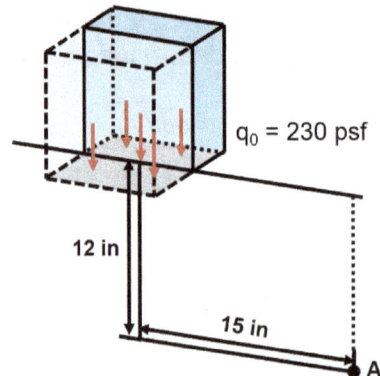

The stress at Point A (q_A) can be determined through the vertical stress distribution contour (Graph B on the right) provided by the reference manual for the square foundation.

footing width $B = 0.5\text{ft} = 0.5\text{ft} \times 12\text{in/ft} = 6\text{in}$

depth of Point A $D_A = 12\text{in} = 2B$

distance of Point A to the center $S_A = 15\text{in} = 2.5B$

Plotting 2.5B as the x-axis and 2B as the y-axis into the square foundation stress distribution graph, the point set on the contour of $0.01q_0$. Therefore, Point A experiences $q_A = 0.01q_0$ stress.

stress at Point A $q_A = 0.01q_0 = 0.01 \times 230\text{psf} = 2.3\text{psf}$

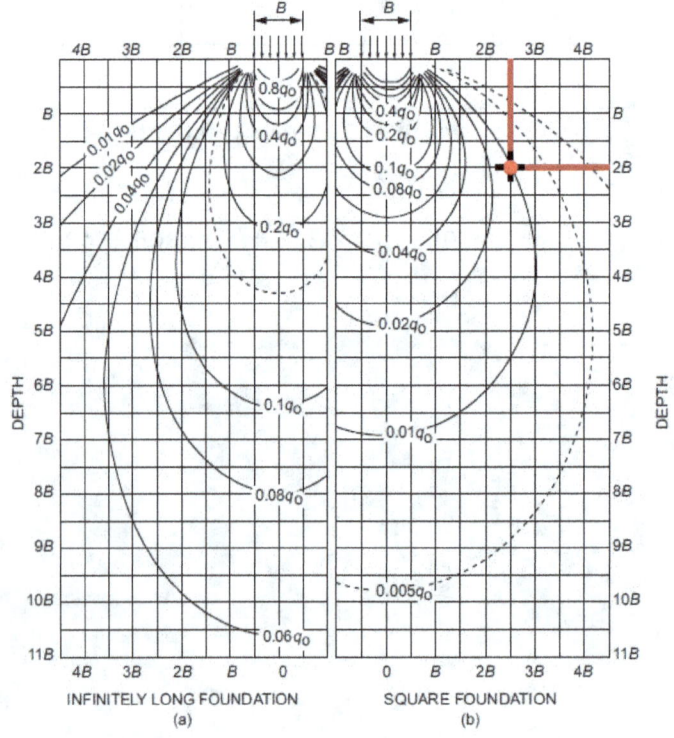

The answer is **(B)**

Solution 12

Analyze the traffic data gathered from a major highway during rush hour to calculate the peak hour factor.

TIME	NUMBER OF VEHICLES
00-07 min	1460
07-15 min	1595
15-22 min	2034
22-30 min	987
30-37 min	1639
37-45 min	2743
45-52 min	1956
52-60 min	1829

The peak-hour factor (PHF) formula is:

$$\text{PHF} = \frac{V}{V_{15} \times 4}$$

hourly volume for an hour of analysis V

$V = \Sigma \text{ vehicles} = 1460 + 1595 + 2034 + 987 + 1639 + 2743 + 1956 + 1829 = 14243$

The maximum 15-minute flow rate within the peak hour (V_{15}) in this specific case happens from 37-53 minutes: $V_{15} = 2743 + 1956 = 4699$

peak-hour factor PHF

$$\text{PHF} = \frac{V}{V_{15} \times 4} = \frac{14243}{4699 \times 4} = 0.76$$

The answer is **(C)**

Solution 13

Bank soil was excavated from a specific borrow pit with a formation level of 32.5 ft before being trucked and compacted. The pit grid and spot heights (units in feet) obtained from a level survey are presented. Assume the soil has a swell factor of 0.2 and a shrinkage of 15%, calculate the final volume of the compacted soil.

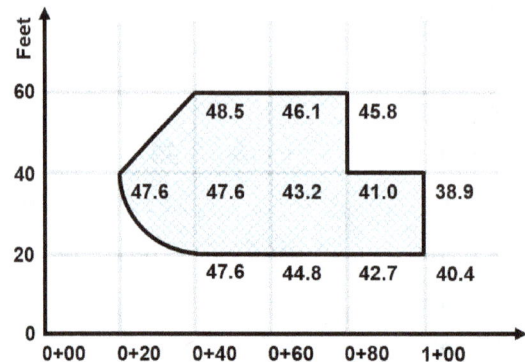

To determine the final volume of the compacted soil, the first step is to calculate the total cut volume of the bank soil (V_{tot}) excavated from the borrow pit with the grid formulas:

The volume of one grid square $V_s = (1/4)(h_1 + h_2 + h_3 + h_4) \times A_s$

The volume of one grid triangle $V_t = (1/3)(h_1 + h_2 + h_3) \times A_t$

The volume of each grid is calculated through the average height (h) times the grid area (A). Please note that a quarter circle is different from a quarter circular cone.

Also, the average height (h) in the formulas refers to the height above the formation level, some height adjustments are needed, take Point X for example:

$h_X = H_X - FL = 48.5ft - 32.5ft = 16.0ft$

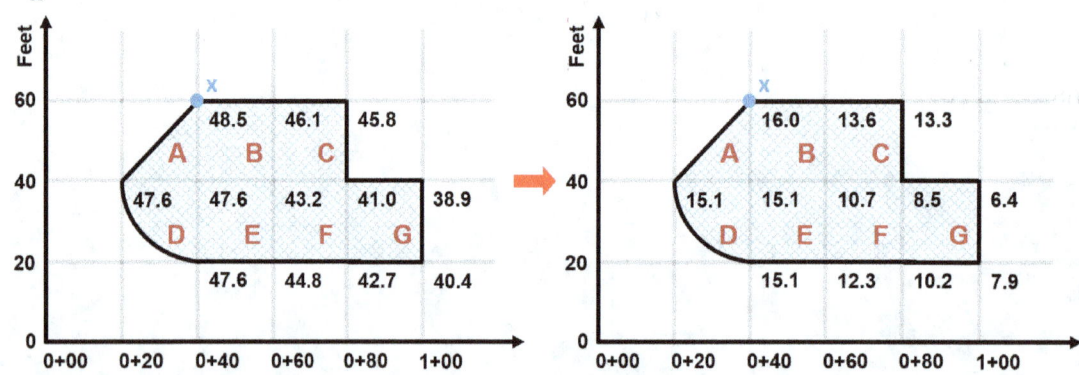

The total volume (V_{tot}) is calculated by adding up the grid volumes, take grid A for example:

$V_A = (1/3)(h_1 + h_2 + h_3) \cdot A_t = (1/3)(15.1ft + 15.1ft + 16.0ft) \times (20ft \times 20ft \div 2) = 3080ft^3$

The volumes for each grid are as follows:

Grid	h_1	h_2	h_3	h_4	Ave h	Area A	Volume V
A	15.1	15.1	16.0	N/A	15.4	200	3080
B	16.0	15.1	10.7	13.6	13.9	400	5540
C	13.6	10.7	8.5	13.3	11.5	400	4610

D	15.1	15.1	15.1	N/A	15.1	100π	4743.8
E	15.1	15.1	12.3	10.7	13.3	400	5320
F	10.7	12.3	10.2	8.5	10.4	400	4170
G	8.5	10.2	7.9	6.4	8.3	400	3300
Total							30763.8

$30763.8 ft^3 = 30763.8 ft^3 \times (1yd/3ft)^3 = 1139.4 yd^3$
Therefore, the total excavated volume $V_{tot} = 30763.8 ft^3 = 1139.4 yd^3$

Since the soil was excavated and swelled (swell factor $S = 0.2$); the volume changed:
swelled volume $V_s = V_{tot} \times (1 + S) = 1139.4 yd^3 \times (1 + 0.2) = 1367.3 yd^3$
Then the soil was compacted and shrank (shrinkage factor $R = 15\%$); the final volume:
compacted volume $V_c = V_s \times (1 - R) = 1367.3 yd^3 \times (1 - 0.15) = 1162.2 yd^3$

The answer is **(D)**

Solution 14

An investor is considering an engineering project with an expected future value of 2 million dollars in 25 years. If the annual interest rate is 4.8%, determine the initial investment (in dollars) the investor needs to make now to achieve this future value.

The present value (P) of the project can be determined through the future value (F):
$$P = F(1 + i)^{-n}$$
future value $F = 2$ million $= 2 \times 10^6$
interest rate per year $i = 4.8\% = 0.048$
number of compounding periods $n = 25 yrs$
present value P
$$P = F(1 + i)^{-n} = 2 \times 10^6 \times (1 + 0.048)^{-25} = 6.2 \times 10^5$$

The answer is **(D)**

Solution 15

River water is passing through a trapezoidal open channel at a flow rate of 0.138 cfs. The cross-section of the channel is shown below. Assuming the water has a kinematic viscosity of 1.2 mm²/s, determine the type of the flow.

The open channel flow type is generally determined through the Reynolds Number (Re), which can be calculated through the formula:

$$Re = \frac{vR_H}{v}$$

The area of the trapezoidal channel $A = (a + b) \times h/2$;
$a = b + 2h/\tan\alpha = 2.4\text{ft} + 2 \times 1.0\text{ft} \div \tan 60° = 3.55\text{ft}$
Therefore, the trapezoidal area $A = (a + b) \times h/2 = (2.4\text{ft} + 3.55\text{ft}) \times 1.0\text{ft} \div 2 = 2.98\text{ft}^2$

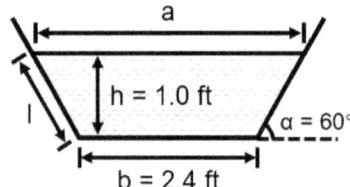

mean flow velocity $v = Q/A = 0.138\text{cfs} \div 2.98\text{ft}^2 = 0.0463\text{ft/s}$
hydraulic radius R_H

$$R_H = \frac{\text{cross} - \text{sectional area}}{\text{wetted perimeter}} = \frac{A}{b + 2l} = \frac{2.98\text{ft}^2}{2.4\text{ft} + 2 \times (1.0\text{ft} \div \sin 60°)} = 0.633\text{ft}$$

kinematic viscosity $v = 1.2\text{mm}^2/\text{s} = 1.2\text{mm}^2/\text{s} \div (304.8\text{mm/ft})^2 = 1.292 \times 10^{-5}\text{ft}^2/\text{s}$
Reynolds Number Re

$$Re = \frac{vR_H}{v} = \frac{0.0463\text{ft/s} \times 0.633\text{ft}}{1.292 \times 10^{-5}\text{ft}^2/\text{s}} = 2268$$

Since the Reynolds Number Re ≥ 2000, the type is the fully turbulent flow.

The answer is **(C)**

Solution 16

A highway soil material has 75 percent of the total passing through the 0.075 mm sieve. The group and plasticity index of the sample are 15.5 and 15 respectively. Determine the AASHTO classification of the material.

The AASHTO classification of the soil sample can be determined with the classification system table in the reference manual.
Based on the information provided, the percent passing the No. 200 sieve (0.075 mm) F = 45; group index GI = 15.5; plasticity index PI = 15; the material might be A-6, A-7-5, or A-7-6.
To further determine the sample classification, the liquid limit (LL) can be calculated through the formula: $GI = (F - 35)[0.2 + 0.005(LL - 40)] + 0.01(F - 15)(PI - 10)$
Plug in all the known information:
$15.5 = (75 - 35)[0.2 + 0.005(LL - 40)] + 0.01(75 - 15)(15 - 10)$ → $LL = 62.5$
Since the sample has a liquid limit (LL) of 62.5 and a plasticity index (PI) of 15, the classification can only be A-2-7 or A-7-5 through the AASHTO plasticity chart;
Therefore, combining all the information, the sample belongs to group A-7-5.

The answer is **(C)**

Solution 17

Following a storm, an average precipitation depth of 2.3 inches has been observed. The area has a curve number of 59, calculate the initial abstraction associated with the rainfall.

The initial abstraction of the rainfall (I_a) can be determined through the formula:

$$\text{Loss (in.)} = P - Q = \frac{(S + I_a) - \frac{I_a^2}{P}}{1 - \frac{I_a}{P} + \frac{S}{P}}$$

precipitation $P = 2.3\text{in}$
maximum basin retention $S = 1000/CN - 10 = 1000 \div 59 - 10 = 6.95\text{in}$
runoff depth (Q) can be calculated with:

$$Q = \frac{(P - 0.2S)^2}{P + 0.8S} = \frac{(2.3\text{in} - 0.2 \times 6.95\text{in})^2}{2.3\text{in} + 0.8 \times 6.95\text{in}} = 0.105\text{in}$$

Therefore, we have initial abstraction I_a:

$$\text{Loss (in.)} = 2.3\text{in} - 0.105\text{in} = \frac{(6.95\text{in} + I_a) - \frac{I_a^2}{2.3\text{in}}}{1 - \frac{I_a}{2.3\text{in}} + \frac{6.95\text{in}}{2.3\text{in}}} \quad \rightarrow \quad I_a = 1.4\text{in}$$

The answer is **(A)**

Solution 18

A concrete circular column with a length of 8 feet is fixed on both sides. Suppose the circular cross-section has a diameter of 20 inches and the column has a modulus of elasticity of 4,400 ksi. What is the critical stress of the column?

The critical stress (σ_{cr}) of the column can be determined through Euler's formula:

$$\sigma_{cr} = \frac{\pi^2 E}{(KL/r)^2}$$

modulus of elasticity $E = 4400 \text{ksi}$
Since both sides of the column are fixed, the effective length factor of column $K = 0.5$
unbraced column length $L = 8\text{ft} = 8\text{ft} \times 12\text{in/ft} = 96\text{in}$
column radius $a = d/2 = 20\text{in} \div 2 = 10\text{in}$

The radius of gyration (r) can be determined through the formula:

$$r = \sqrt{I/A}$$

moment of initial $I = \pi a^4/4 = 3.14 \times (10\text{in})^4 \div 4 = 7854 \text{in}^4$
area of the circle $A = \pi a^2 = 3.14 \times (10\text{in})^2 = 314 \text{in}^2$
radius of gyration r

$$r = \sqrt{I/A} = \sqrt{7854 \text{in}^4 \div 314 \text{in}^2} = 5\text{in}$$

critical stress σ_{cr}

$$\sigma_{cr} = \frac{\pi^2 E}{(KL/r)^2} = \frac{3.14^2 \times 4400 \text{ksi}}{(0.5 \times 96\text{in} \div 5\text{in})^2} = 471 \text{ksi}$$

The answer is **(B)**

Solution 19
Draw the shear diagram for the beam with the following loading conditions:

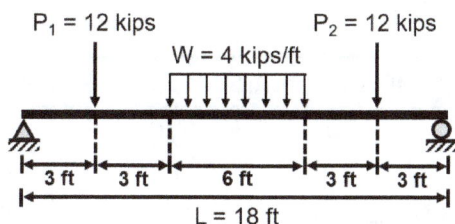

The first step to draw the shear diagram is to calculate the support reaction forces:

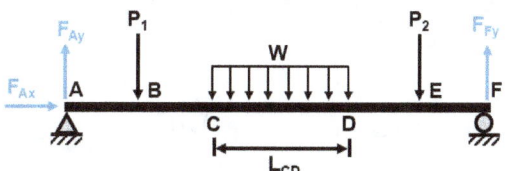

For the beam, the sum of forces (y-axis) is 0:
$F_{Ay} + F_{Fy} = P_1 + WL_{CD} + P_2 = 12\text{kips} + 4\text{kips/ft} \times 6\text{ft} + 12\text{kips} = 48\text{kips}$
Since the forces P_1, W, P_2 are symmetrical to the A and F sides, $F_{Ay} = F_{Fy} = 24\text{kips}$

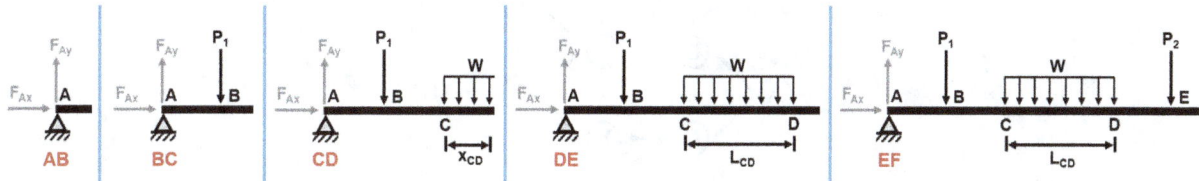

To calculate the shear force (V), different sections of the beam are considered separately:
For beam section AB, the sum of forces on the y-axis $\sum F_{ABy} = 0$, all units in kips
$\therefore F_{Ay} - V_{AB} = 24 - V_{AB} = 0$ $\qquad V_{AB} = 24$
For beam section BC, $\sum F_{BCy} = 0$
$\therefore F_{Ay} - P_1 - V_{BC} = 24 - 12 - V_{BC} = 0$ $\qquad V_{BC} = 12$
For beam section CD, $\sum F_{CDy} = 0$
$\therefore F_{Ay} - P_1 - WX_{CD} - V_{CD} = 24 - 12 - 4 \times X_{CD} - V_{CD} = 0$ $\qquad V_{CD} = 12 - 4X_{CD}$
For beam section DE, $\sum F_{DEy} = 0$
$\therefore F_{Ay} - P_1 - WL_{CD} - V_{DE} = 24 - 12 - 4 \times 6 - V_{DE} = 0$ $\qquad V_{DE} = -12$
For beam section EF, $\sum F_{EFy} = 0$
$\therefore F_{Ay} - P_1 - WL_{CD} - P_2 - \text{shear}_{EF} = 24 - 12 - 4 \times 6 - 12 - V_{EF} = 0$ $\qquad V_{EF} = -24$

The answer is

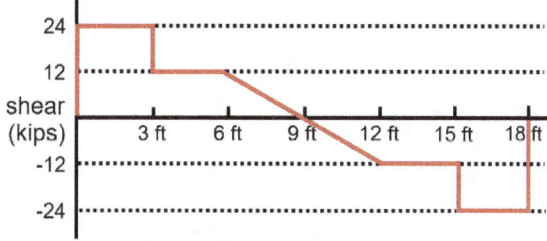

Solution 20

Determine the minimum time necessary to complete the entire industrial construction project based on the scheduled activity data.

ACTIVITY	PREDECESSOR	SUCCESSOR	DURATION IN DAYS
A	-	B,C	4
B	A	D	7
C	A	E	2
D	B,G	F	3
E	C	G	5
F	D		8
G	E	D,H	1
H	G	-	6

Based on the activity data, a path diagram can be drawn:

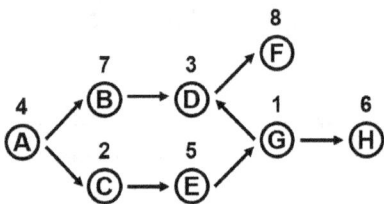

To determine the minimum time necessary to complete the entire project (T), one must follow the longest effective path (in time) from start to finish.

The paths from start to finish include 1. A-B-D-F; 2. A-C-E-G-D-F; 3. A-C-E-G-H
The time required for each of them are:
$T_1 = 4 days + 7 days + 3 days + 8 days = 22 days$
$T_2 = 4 days + 2 days + 5 days + 1 days + 3 days + 8 days = 23 days$
$T_3 = 4 days + 2 days + 5 days + 1 days + 6 days = 18 days$

Therefore, the path A-C-E-G-D-F is the critical path and the minimum number of days to complete the project is 23 days.

The answer is **(D)**

Solution 21

A retaining wall shown as follows is designed to support a cohesive backfill with a unit weight of 110 lb/ft³. Assuming an angle of internal friction of 30°, determine the active resultant force induced on the wall at a depth of 3.2 feet.

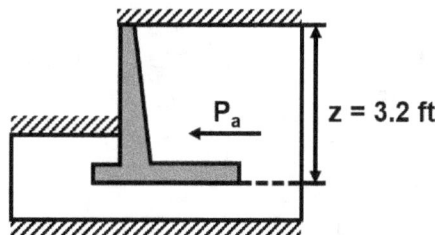

The active resultant force (P_a) can be calculated through the formula: $P_a = K_a \gamma z^2 / 2$
coefficient of active earth pressure K_a

$$K_a = \frac{1 - \sin\phi'}{1 + \sin\phi'} = \tan^2\left(45 - \frac{\phi'}{2}\right) = \tan^2\left(45 - \frac{30°}{2}\right) = \frac{1}{3}$$

Given depth $z = 3.2$ ft, soil unit weight $\gamma = 110$ lb/ft³;
active resultant force P_a

$$P_a = \frac{K_a \gamma z^2}{2} = \frac{1/3 \times 110 \text{lb/ft}^3 \times (3.2\text{ft})^2}{2} = 187.7 \text{lb/ft}$$

The answer is **(A)**

Solution 22

A wastewater treatment cylinder sedimentation tank has a diameter of 30 feet and a depth of 12 feet. Suppose the influent flow rate is 500,000 gallons per day, calculate the detention time in hours for the sedimentation tank.

The hydraulic detention time (τ) can be determined through the formula: $\tau = V/Q$
tank volume $V = \pi d^2 h / 4 = 3.14 \times (30\text{ft})^2 \times 12\text{ft} \div 4 = 8478 \text{ft}^3$
influent flow rate (which equals the effluent flow rate) Q

$$Q = 5 \times 10^5 \frac{\text{gallon}}{\text{day}} = 5 \times 10^5 \frac{\text{gallon}}{\text{day}} \times 0.134 \frac{\text{ft}^3}{\text{gallon}} \times \frac{1}{24} \frac{\text{day}}{\text{hr}} = 2791.67 \text{ft}^3/\text{hr}$$

hydraulic detention time τ

$$\tau = \frac{V}{Q} = \frac{8478 \text{ft}^3}{2791.67 \text{ft}^3/\text{hr}} = 3.04 \text{hr}$$

The answer is **(C)**

Solution 23
A clay layer with a thickness of 8 feet is subjected to an effective stress of 1500 psf. Assume the clay has a compression index of 0.07 and an initial void ratio of 1.18. What is the consolidation settlement when an additional 1000 psf effective stress is loaded vertically to the top of the layer?

To calculate the consolidation settlement (S_c) of the clay, the following formula can be used:
$$S_c = \sum_1^n \frac{H_0}{1+e_0}(C_r \log_{10}\frac{p_c}{p_0} + C_c \log_{10}\frac{p_f}{p_c})$$
Since only one soil layer is considered, $n = 1$
thickness of compressible layer $H_0 = 8ft = 96in$;
initial overburden pressure $p_0 = p_c = 1500psf$
final overburden pressure $p_f = p_0 + \Delta p = 1500psf + 1000psf = 2500psf$
consolidation settlement S_c
$$S_c = \sum_1^n \frac{H_0}{1+e_0}(C_r \log_{10}\frac{p_c}{p_0} + C_c \log_{10}\frac{p_f}{p_c}) = \sum_1^1 \frac{96in}{1+1.18} \times (0 + 0.07 \times \log_{10}\frac{2500psf}{1500psf}) = 0.684in$$

The answer is **(D)**

Solution 24
A 0.8-acre drainage pond is designed to collect rainwater from storm events with a retention time of 48 hours. Historical records indicate that the largest local storm events produced a rainfall intensity of 4.7 in/hr over an area of 3.3 acres. Assume the area has a runoff coefficient of 0.25. Determine the minimum depth required for the pond to hold the largest events.

The minimum depth (d) of the pond can be determined through the relationship between the pond volume and the area $d = V/A$, since volume (V) can also be represented by the hydraulic retention time formula $V = Q\tau$, we have $d = V/A = Q\tau/A$;

According to the rational rainfall-runoff method, flow rate $Q = CIA_{tot}$, given runoff coefficient $C = 0.25$; rainfall intensity $I = 4.7in/hr$; total drainage area $A_{tot} = 3.3acre$; flow rate Q
$$Q = CIA_{tot} = 0.25 \times 4.7in/hr \times 3.3acre = 3.88 acre \cdot in/hr$$
Given retention time $\tau = 48hr$; retention pond area $A = 0.8acre$;
minimum depth $d = Q\tau/A = 3.88 acre \cdot in/hr \times 48hr \div (0.8acre) = 232.8in = 19.4ft$

The answer is **(C)**

Solution 25

A construction site requires dewatering to regulate the groundwater table with a 6-inch-diameter penetrating well. Detailed data regarding the unconfined aquifer is provided below. Assume the hydraulic conductivity estimated from the slug test is 36×10⁻⁴ cm/s. Determine the flow rate from the well given all the information provided.

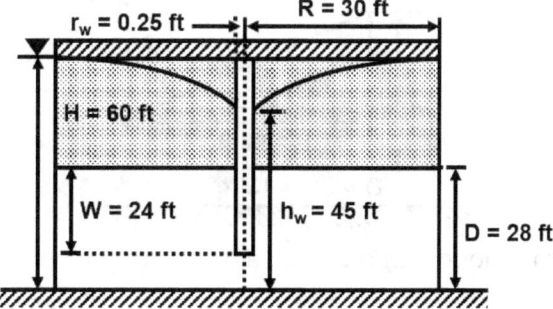

Flow (Q_{wp}) from the partially penetrating well can be calculated through:

$$Q_{wp} = \frac{2\pi k D(H - h_w)G}{\ln(R/r_w)}$$

hydraulic conductivity $k = 36 \times 10^{-4} cm/s = 36 \times 10^{-4} cm/s \times 0.0328 ft/cm = 1.1811 \times 10^{-4} ft/s$

well radius $r_w = d_w/2 = 6in/2 = 3in = 0.25ft$

partial to full ratio G

$$G = \frac{W}{D}\left(1 + 7\sqrt{\frac{r_w}{2W}} \cos\frac{\pi w/D}{2}\right) = \frac{24ft}{28ft}\left(1 + 7\sqrt{\frac{0.25ft}{2 \times 24ft}} \cos\frac{180° \times 24ft \div 28ft}{2}\right) = 0.9535$$

Please note that the π above stands for radians instead of numbers, $\pi = 180°$.

flow rate Q_{wp}

$$Q_{wp} = \frac{2\pi k D(H - h_w)G}{\ln(R/r_w)} = \frac{2 \times \pi \times 1.1811 \times 10^{-4} ft/s \times 28ft \times (60ft - 45ft) \times 0.9535}{\ln(30ft \div 0.25ft)} = 0.062 ft^3/s$$

The answer is **(B)**

Solution 26

A geotechnical engineer is working on a project constructing a slope for the highway embankment. The slope is comprised of soil with a unit weight of 125 lb/ft³ and a cohesion of 50 kPa. Detailed information on the slope scale is provided below. Determine the factor of safety for this slope against sliding.

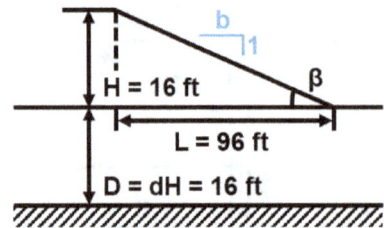

The factor of safety (F) for this slope can be determined through the formula:

$$F = N_0 \frac{c}{\gamma H}$$

stability number (N_0) can be determined through the slope stability chart in the reference manual:
∵ D = dH   d = D/H = 16ft ÷ 16ft = 1
∵ cotβ = L/H = 96ft ÷ 16ft = 6

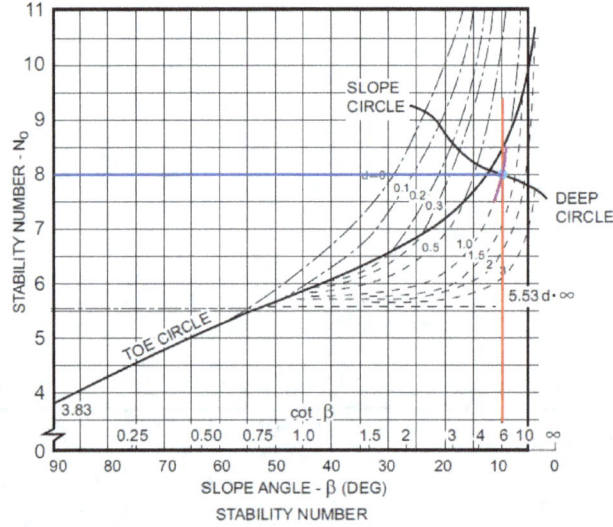

From the chart, the stability number $N_0 = 8$
soil cohesion $c = 50\text{kPa} = 50\text{kPa} \times 20.89\text{psf/kPa} = 1044.3\text{psf}$
soil unit weight $\gamma = 125 \text{lbf/ft}^3$; slope height $H = 16\text{ft}$
factor of safety F

$$F = N_0 \frac{c}{\gamma H} = 8 \times \frac{1044.3\text{psf}}{125\text{lbf/ft}^3 \times 16\text{ft}} = 4.2$$

The answer is **(D)**

Solution 27

A 10" ×10" concrete column is subjected to an eccentric load of 64 kips including self-weight with an eccentricity of 2 inches to the right axis. Determine the maximum column compressive stress caused by this loading.

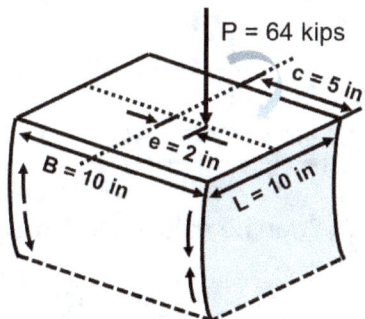

The compressive stress (σ_c) in a column with eccentric loading includes two different parts: the concentric loading stress and the internal bending stress, compressive stress $\sigma_c = P/A \pm Mc/I$, since we are calculating the maximum compressive stress, the formula should be:

$$\sigma_{cM} = \frac{P}{A} + \frac{Mc}{I}$$

loading stress $P = 64 \text{kips}$
cross-section area $A = BL = 10\text{in} \times 10\text{in} = 100\text{in}^2$
eccentric load moment $M = P \cdot e = 64\text{kips} \times 2\text{in} = 128\text{kip} \cdot \text{in}$
distance from the neutral axis to the outermost fiber of a symmetrical beam section $c = 5\text{in}$
moment of inertia $I = LB^3/12 = 10\text{in} \times (10\text{in})^3 \div 12 = 833.3\text{in}^4$
maximum compressive stress σ_{cM}

$$\sigma_{cM} = \frac{P}{A} + \frac{Mc}{I} = \frac{64\text{kips}}{100\text{in}^2} + \frac{128\text{kip} \cdot \text{in} \times 5\text{in}}{833.3\text{in}^4} = 1.41\text{ksi}$$

The answer is **(C)**

Solution 28

A 10" ×10" concrete column is subjected to an eccentric load of 64 kips with an eccentricity of 2 inches to the right axis. Calculate the maximum bearing pressure on the column footing.

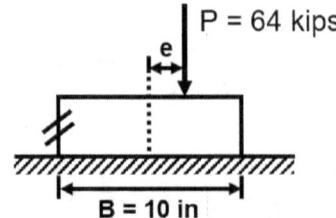

The maximum bearing pressure (q_{max}) can be calculated through the formula:

$$q_{max} = \begin{cases} \dfrac{P}{BL}\left(1+\dfrac{6e}{B}\right), & \text{for } e < \dfrac{B}{6} \\ \dfrac{4P}{3L(B-2e)}, & \text{for } e > \dfrac{B}{6} \end{cases}$$

Since the eccentricity $e = 2\text{in}$ and the footing width $B = 10\text{in}$, $(e = 2\text{in}) > (B/6 = 10/6\text{in})$; maximum bearing pressure q_{max}

$$q_{max} = \frac{4P}{3L(B-2e)} = \frac{4 \times 64\text{kips}}{3 \times 10\text{in} \times (10\text{in} - 2 \times 2\text{in})} = 1.42\text{ksi}$$

The answer is **(B)**

Solution 29

The water level in an aquifer is observed to decline from 129.6 inches to 129.3 inches over a distance of 370 feet. Suppose the aquifer has a specific discharge of 5.7 ft/day and a porosity of 0.4, what is the hydraulic conductivity and average seepage velocity of the aquifer?

The hydraulic conductivity (k) of the aquifer can be determined through the formula:

$$q = -k\frac{dh}{dx} \quad \rightarrow \quad k = -q\frac{dx}{dh}$$

specific discharge $q = 5.7\text{ft/day} = 0.2375\text{ft/h} = 6.6 \times 10^{-5}\text{ft/s}$
elevation change $dh = h_2 - h_1 = 129.3\text{in} - 129.6\text{in} = -0.3\text{in} = -0.025\text{ft}$
hydraulic conductivity k

$$k = -q\frac{dx}{dh} = -6.6 \times 10^{-5}\text{ft/s} \times \frac{370\text{ft}}{-0.025\text{ft}} = 0.98\text{ft/s}$$

The average seepage velocity (v) can be determined through the formula:

$$v = \frac{q}{S_y} = \frac{0.2375\text{ft/hr}}{0.4} = 0.59\text{ft/hr}$$

The answer is **0.98 and 0.59**

Solution 30

A mobile crane with a self-weight of 110 kips and a telescoping boom weight of 10 kips is positioned on an industrial site with dimensions shown below. Given a total contact area of 25 ft², determine the average ground pressure when the crane reaches its maximum load capacity before tips over.

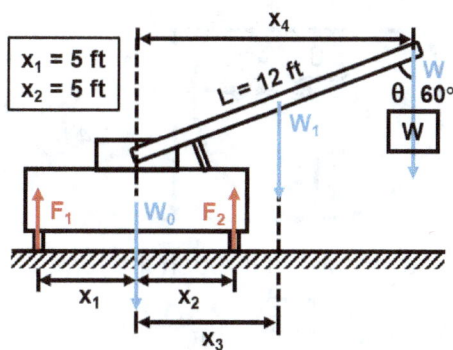

To calculate the ground pressure (p), the maximum loading and supporting force need to be determined first; before the crane tips over, the sum of moments and forces remains 0.
left crane foot moment: $\sum M = -W_0 x_1 + F_2(x_1 + x_2) - W_1(x_1 + x_3) - W(x_1 + x_4) = 0$
Y-axis sum of force: $\sum F = F_1 + F_2 - W_0 - W_1 - W = 0$
PS: the moment balance equation may be subject to changes based on the reference selected.

lengths $x_1 = x_2 = 5ft$, $x_3 = L \cdot \sin\theta/2 = 12ft \times \sin60° \div 2 = 5.2ft$, $x_4 = 2x_3 = 2 \times 5.2ft = 10.4ft$;
Before overturning, the left footing is at the critical point leaving the ground, therefore, $F_1 = 0$;

Taking everything into original equations:
$\begin{cases} -110\text{kips} \times 5ft + F_2 \times (5ft + 5ft) - 10\text{kips} \times (5ft + 5.2ft) - W \times (5ft + 10.4ft) = 0 \\ 0 + F_2 - 110\text{kips} - 10\text{kips} - W = 0 \end{cases}$

$\begin{cases} W = 101.5\text{kips} \\ F_2 = 221.5\text{kips} \end{cases}$

ground pressure $p = F_2/A = 221.5\text{kips} \div 25ft^2 = 8.9\text{kips}/ft^2$

The answer is **(C)**

Solution 31

In a 1.6 ft radius circular pipe channel, water flows at a depth of 2.4 ft as shown in the figure. Given a Manning's roughness coefficient of 0.014 for the pipe material and a longitudinal slope of 0.25%, calculate the flow rate of water through the pipe.

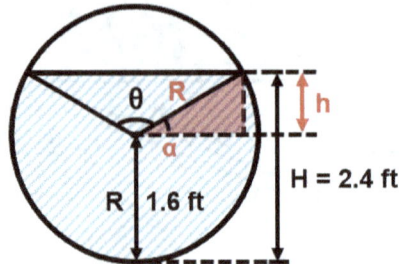

The flow rate (Q) in the pipe can be calculated through the formula:

$$Q = \frac{1.486}{n} A R_H^{2/3} S^{1/2}$$

Manning's roughness coefficient $n = 0.014$; slope of the energy grade line $S = 0.25\%$;
angle $\alpha = \sin^{-1}(h/R) = \sin^{-1}[(H-R)/R] = \sin^{-1}[(2.4\text{ft} - 1.6\text{ft}) \div 1.6\text{ft}] = 30°$
angle $\theta = 180° - 2\alpha = 180° - 2 \times 30° = 120°$
cross-sectional area of flow (A) is the addition of the 2/3 circle and a triangle:

$$A = \pi R^2 \times \frac{360° - \theta}{360°} + h \cdot \frac{h}{\tan\alpha} = 3.14 \times (1.6\text{ft})^2 \times \frac{360° - 120°}{360°} + 0.8\text{ft} \cdot \frac{0.8\text{ft}}{\tan 30°} = 6.47\text{ft}^2$$

hydraulic radius R_H

$$R_H = \frac{\text{cross} - \text{sectional area}}{\text{wetted perimeter}} = \frac{A}{2\pi R \times \frac{(360° - \theta)}{360°}} = \frac{6.47\text{ft}^2}{2 \times 3.14 \times 1.6\text{ft} \times \frac{360° - 120°}{360°}} = 0.966\text{ft}$$

Please note that the non-filled pipe area and the hydraulic radius can also be calculated with formulas provided in the reference manual (angle is in radians):
$A = 1/8(\varphi - \sin\varphi)d_0^2 \qquad R_H = 1/4(1 - \sin\varphi/\varphi)d_0$
$\varphi = 360° - 120° = 4/3\pi \qquad d_0 = 2R = 3.2\text{ft}$

flow rate Q

$$Q = \frac{1.486}{n} A R_H^{2/3} S^{1/2} = \frac{1.486}{0.014} \times 6.47 \times (0.966)^{2/3} \times (0.25\%)^{1/2} \text{cfs} = 33.6\text{cfs}$$

The answer is **(A)**

Solution 32

The survey leveling data gathered for a construction site is given below. Determine the ground elevation of point TP2.

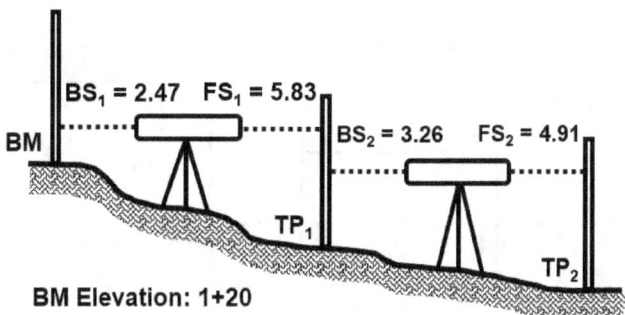

BM Elevation: 1+20

The elevation of turning points has the following relationships:
$BM + BS = HI \quad TP = HI - FS \quad \rightarrow \quad BM + BS = TP + FS$

In this question:
$\begin{cases} BM + BS_1 = TP_1 + FS_1 \\ TP_1 + BS_2 = TP_2 + FS_2 \end{cases} \rightarrow \begin{cases} 120' + 2.47' = TP_1 + 5.83' \\ TP_1 + 3.26' = TP_2 + 4.91' \end{cases}$

Given the data provided, $TP_1 = 116.64'$, $TP_2 = 114.99'$

The answer is **(C)**

Solution 33

A city with a population of 2.1 million people is growing exponentially at a rate of 2.18%. According to the municipal water distribution on fire protection demands, determine the required fire flow in 8 years.

Since the population is growing exponentially, in 8 years, the city population (P_t) will change to:
$$P_t = P_0 \cdot e^{k\Delta t} = 2.1 \times 10^6 \times e^{2.18\% \times 8} = 2500117 = 2500k$$

The required fire flow Q
$$Q = 1020\sqrt{P}(1 - 0.01\sqrt{P}) = 1020\sqrt{2500}(1 - 0.01\sqrt{2500}) = 25500 \text{gal/min} = 56.81 \text{cfs}$$

Please note that the formula above applies population (P) in thousands.

The answer is **(B)**

Solution 34

A table framing is shown below with supporting load information provided. Determine the total load on the 2-feet-long BII column.

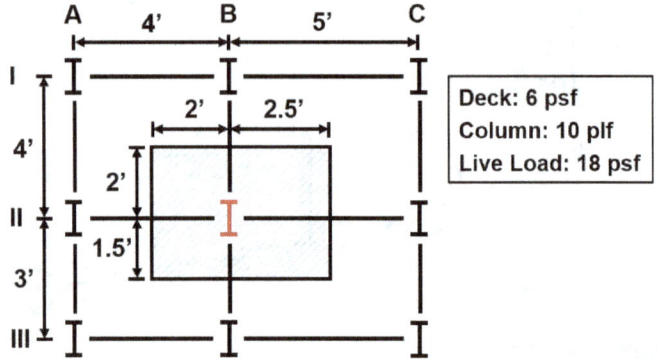

The total load (L_T) on the BII column contains the total surface load (L_S) and the column weight load (L_C): $L_T = L_S + L_C$; the total surface load (L_S) can be calculated through total surface pressure (P_S) times the tributary area (A_t): $L_S = P_S \cdot A_t$; and the column weight load (L_C) can be calculated by multiplying the column length weight (W_C) and the column length (x_C): $L_C = W_C \cdot x_C$

total surface pressure $P_S = P_{deck} + P_{live\ load} = 6psf + 18psf = 24psf$
tributary area A_t

$$A_t = \left(\frac{x_{AB}}{2} + \frac{x_{BC}}{2}\right) \cdot \left(\frac{x_{I-II}}{2} + \frac{x_{II-III}}{2}\right) = \left(\frac{4ft}{2} + \frac{5ft}{2}\right) \times \left(\frac{4ft}{2} + \frac{3ft}{2}\right) = 15.75ft^2$$

total surface load $L_S = P_S \cdot A_t = 24psf \times 15.75ft^2 = 378lb$

column weight load $L_C = W_C \cdot x_C = 10plf \times 2ft = 20lb$
total load $L_T = L_S + L_C = 378lb + 20lb = 398lb$

The answer is **(D)**

Solution 35

A fine sand layer and shallow wall foundation are shown below. With a combined weight of 85,000 lbf for the wall and the rigid circular footing, determine the surface vertical settlement.

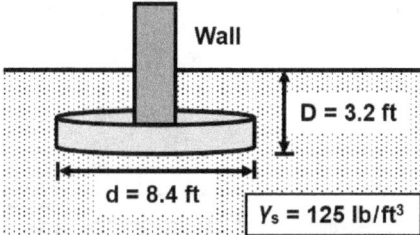

The vertical settlement (δ_v) can be determined through the formula: $\delta_v = C_d \Delta p B_f (1 - v^2)/E_m$
Since the loaded area is a rigid circle, the shape and rigidity factor $C_d = 0.79$; the footing diameter $B_f = d = 8.4 \text{ft}$; Given the elastic constant of soil (reference manual), the fine sand Poisson's Ratio $v = 0.25$; the elastic modulus $E_m = 180 \text{tsf} = 180 \text{tsf} \times 2000 \text{lbf/ton} = 3.6 \times 10^5 \text{psf}$;
footing area $S = \pi d^2/4 = 3.14 \times (8.4 \text{ft})^2 \div 4 = 55.4 \text{ft}^2$
footing stress change $\Delta p = F/S - \gamma_s \cdot D = 85000 \text{lbf} \div 55.4 \text{ft}^2 - 125 \text{lb/ft}^3 \times 3.2 \text{ft} = 1134.3 \text{psf}$
vertical settlement δ_v

$$\delta_v = \frac{C_d \Delta p B_f (1 - v^2)}{E_m} = \frac{0.79 \times 1134.3 \text{psf} \times 8.4 \text{ft} \times (1 - 0.25^2)}{3.6 \times 10^5 \text{psf}} = 0.0196 \text{ft} = 0.235 \text{in}$$

The answer is **(B)**

Solution 36

A water distribution system utilizes a 20-meter-long plastic circular pipe with a diameter of 0.6 m to convey water. The flow velocity through the pipe is 0.25 cm/s. Calculate the head loss due to friction in the pipe using the Hazen-Williams equation.

The head loss of the circular pipe (h_f) can be calculated through the formula:

$$h_f = \frac{4.73 L}{C^{1.852} D^{4.87}} Q^{1.852}$$

pipe length of head loss $L = 20\text{m} = 20\text{m} \times 3.28 \text{ft/m} = 65.6 \text{ft}$
pipe diameter $D = 0.6\text{m} = 1.97 \text{ft}$
flow $Q = v \cdot A = v \cdot (\pi D^2/4) = 0.25 \times 10^{-2} \text{m/s} \times 3.28 \text{ft/m} \times [3.14 \times (1.97\text{ft})^2 \div 4] = 0.025 \text{cfs}$
Hazen-Williams coefficient for plastic pipe $C = 150$
head loss h_f

$$h_f = \frac{4.73 L}{C^{1.852} D^{4.87}} Q^{1.852} = \frac{4.73 \times 65.6 \text{ft}}{150^{1.852} \times 1.97^{4.87}} 0.025^{1.852} = 1.15 \times 10^{-6} \text{ft}$$

The answer is **(B)**

Solution 37

The house construction involves the following tasks shown with a table and a node notation. Suppose two groups of workers of 4 (workers not transferrable) are undertaking the entire construction project, and the cost per day per worker is $200. Estimate the minimum duration and the cost of the project.

ORDER	TASK	DURATION (DAYS)
A	Site Preparation	10
B	Foundation laying	15
C	Construction	30
D	Interior finishing	20
E	Landscaping	10

A+B: Since tasks A and B start together, two groups can work separately on different tasks, for the first 10-day workload, each group needs 20 days to finish (the duration in the table per task is the time required for two groups working together). After 20 days, task B remains a 5-day workload, two groups can work together to complete the rest, which takes another 5 days. As a result, tasks A and B take 25 days in total to finish.

C+D: Since tasks C and D are completed at the same time and task C has a 10-day workload more than task D, the two groups can work together for the first 10-day workload, which takes 10 days. For the remainder 20-day workload for both tasks C and D, the two groups can work separately, which takes 40 days to finish. Therefore, tasks C and D take 50 days in total to finish.

E: For task E, two groups can work together for the 10-day workload that takes 10 days.

As a result, the minimum duration of the project $D = 25\text{days} + 50\text{days} + 10\text{days} = 85\text{days}$

Since all the workers work consecutively for 85 days, the cost of the project E

$$E = 2\text{group} \times \frac{4\text{worker}}{\text{group}} \times \frac{\$200}{\text{worker} \cdot \text{day}} \times 85\text{day} = \$136000$$

minimum cost of the project $E = \$136000$

The answer is **(D)**

Solution 38

A water remediation project employs a specific adsorbent to reduce PFAS contamination in an open stream, achieving an adsorption efficiency of 80%. If the initial PFAS concentration is 10 mg/L and the water is treated at a flow rate of 500 gallons per day, calculate the total mass of PFAS removed over 15 days.

The total mass of PFAS removed (M) can be calculated by the PFAS concentration decrement times the treated water volume: $M = \Delta C \cdot V = (C_0 - C) \cdot V$
total volume $V = Q \cdot t = 500 \text{gallon/day} \times 15 \text{day} = 7500 \text{gallon}$
ending concentration $C = C_0(1 - e) = 10 \text{mg/L} \times (1 - 80\%) = 2 \text{mg/L}$
concentration decrement $\Delta C = C_0 - C = 10 \text{mg/L} - 2 \text{mg/L} = 8 \text{mg/L} = 6.68 \times 10^{-5} \text{lb/gallon}$
total mass of PFAS removed $M = \Delta C \cdot V = 6.68 \times 10^{-5} \text{lb/gallon} \times 7500 \text{gallon} = 0.50 \text{lb}$

The answer is **(C)**

Solution 39

8 solid pipes shown below are required to be installed for a new construction site. Suppose the inner diameter of the steel pipe is 10", and the costs of the steel and the filled concrete are 28 LF and 110 CY. Determine the total cost of the pipes needed.

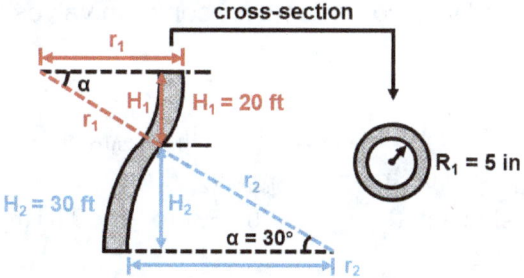

The total cost (C_{tot}) for 8 pipes can be divided into steel cost and concrete cost: $C_{tot} = 8(C_s + C_c)$, where steel cost $C_s = L_s \cdot 28 \text{ft}^{-1}$ and concrete cost $C_c = V_c \cdot 110 \text{yd}^{-3}$
pipe length L_s

$$L_s = 2\pi(r_1 + r_2) \cdot \left(\frac{\alpha}{360°}\right) = \frac{2\pi(H_1 + H_2)}{\sin\alpha} \cdot \left(\frac{\alpha}{360°}\right) = \frac{2\pi(20\text{ft} + 30\text{ft})}{\sin 30°} \cdot \frac{30°}{360°} = 52.36 \text{ft}$$

inner pipe volume $V_c = \pi R_1^2 L_s = 3.14 \times [10\text{in} \div (12\text{in/ft}) \div 2]^2 \times 52.36\text{ft} = 28.54\text{ft}^3 = 1.06\text{yd}^3$
pipes total cost $C_{tot} = 8(C_s + C_c) = 8 \times (52.36\text{ft} \times 28\text{ft}^{-1} + 1.06\text{yd}^3 \times 110\text{yd}^{-3}) = 12661$

The answer is **(C)**

Solution 40

Certain components of a municipal water system are shown below. Tap water passed through a fully opened gate valve into a narrower tube, navigated two short-radius elbows, a pressure regulating valve, and exited the tubing. Given an inlet flow rate of 5 GPM and a total minor head loss of 0.2 meters, determine the pressure regulating valve head loss coefficient.

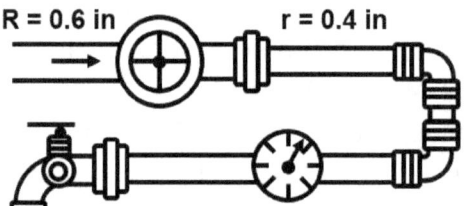

The minor head loss (h_f) can be calculated through the formula:

$$h_f = C \frac{v^2}{2g}$$

total head loss $h_t = h_{f,gate} + 2 \times h_{f,elbow} + h_{f,pressure} + h_{f,exit} = 0.2m$

flow velocity original v_1

$$v_1 = \frac{Q}{A_1} = 5\frac{\text{gallon}}{\text{min}} \times 0.13368 \frac{\text{ft}^3}{\text{gallon}} \div \left[\pi \left(\frac{0.6\text{in}}{12\text{in/ft}}\right)^2\right] = 85.1 \text{ft/min} = 0.432 \text{m/s}$$

flow velocity in narrower tubes v_2

$$v_2 = \frac{Q}{A_2} = 5\frac{\text{gallon}}{\text{min}} \times 0.13368 \frac{\text{ft}^3}{\text{gallon}} \div \left[\pi \left(\frac{0.4\text{in}}{12\text{in/ft}}\right)^2\right] = 191.5 \text{ft/min} = 0.973 \text{m/s}$$

gate head loss $h_{f,gate} = C_{gate} v_1^2 / 2g = 0.2 \times (0.432 \text{m/s})^2 \div (2 \times 9.81 \text{m/s}^2) = 1.90 \times 10^{-3} \text{m}$

elbow head loss $h_{f,elbow} = C_{elbow} v_2^2 / 2g = 0.9 \times (0.973 \text{m/s})^2 \div (2 \times 9.81 \text{m/s}^2) = 0.04343 \text{m}$

exit head loss $h_{f,exit} = C_{exit} v_2^2 / 2g = 1.0 \times (0.973 \text{m/s})^2 \div (2 \times 9.81 \text{m/s}^2) = 0.04825 \text{m}$

Please note that the minor head loss coefficients for common valves used above can be found in the reference manual.

total head loss $h_t = 1.90 \times 10^{-3} \text{m} + 2 \times 0.04343 \text{m} + h_{f,pressure} + 0.04825 \text{m} = 0.2 \text{m}$

pressure head loss $h_{f,pressure} = 0.063 \text{m} = C_{pressure} v_2^2 / 2g$

pressure valve head loss coefficient $C_{pressure} = 0.063 \text{m} \times 2 \times 9.81 \text{m/s}^2 \div (0.973 \text{m/s})^2 = 1.31$

The answer is **(A)**

Water Depth Questions

40 Questions

4 hours

1. A triangular channel shown below has a flow rate of 125 ft³/sec and an average depth of 3 meters. What is the approximate critical depth of the channel?

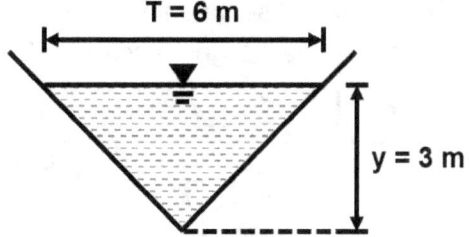

(A) 2.38 ft
(B) 3.96 ft
(C) 4.26 ft
(D) 7.86 ft

2. In an area characterized by commercial and business areas situated on shallow loess sandy loam, a watershed with a length of 20 meters and a slope of 0.16% is utilized for stormwater hydrological management. Calculate the time of concentration for this watershed.

(A) 35 min
(B) 50 min
(C) 58 min
(D) 76 min

3. A simplified city wastewater treatment facility for phosphorous (P) removal is shown below. Assuming the influent has a flow rate of 750.0 GPM with a P concentration of 6.3 mg/L, and the waste stream contains 184.6 mg/L of sediment P. Determine the effluent flow rate required when the P concentration meets the discharge standard of 0.1 mg/L.

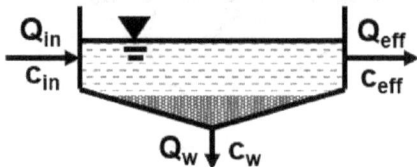

(A) 750.0 GPM
(B) 749.6 GPM
(C) 724.8 GPM
(D) 724.4 GPM

4. A triangular unit runoff hydrograph for a 160-acre watershed is shown below. Suppose the soil infiltration can be ignored, determine the peak discharge per centimeter of precipitation.

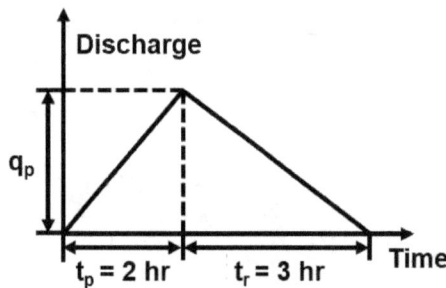

(A) 25.2 cfs
(B) 21.3 cfs
(C) 26.6 cfs
(D) 32.0 cfs

5. A rectangular primary clarifier is 36 feet long and 24 feet 6 inches wide. The average influent flow rate to the clarifier is 490 GPM with a total suspended solids (TSS) concentration of 1.2 g/L. Considering the typical removal efficiency, what would be the effluent TSS concentration?

   (A) 768 mg/L
   (B) 696 mg/L
   (C) 552 mg/L
   (D) 432 mg/L

6. Diesel oil is pumped from the storage tank to the ground gas station 10 feet above. Assuming the outlet oil pressure is 30 psi and the oil flow rate remains 8.3 GPM (nozzle diameter 0.86 inches), determine the pump break horsepower when the pump has an efficiency of 95%.

   (A) 0.02 hp
   (B) 0.11 hp
   (C) 0.17 hp
   (D) 0.20 hp

7. Water is discharged from a sluice gate shown below. A hydraulic jump is observed to occur at Point A where the flow velocity is 18 ft/s at the depth of 2.4 meters. Determine the hydraulic jump classification from Point A to B.

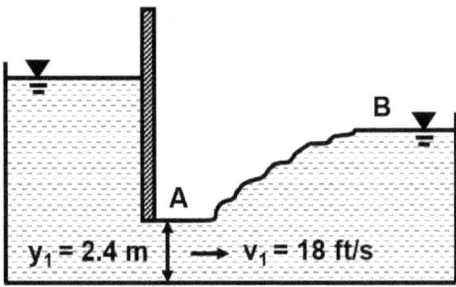

(A) Oscillating Jump
(B) Undular Jump
(C) Steady Jump
(D) Weak Jump

8. In a given year, a natural lake experienced a surface inflow of 2,000 m³ and a drainage outflow of 1,600 m³. The average documented monthly precipitation and evaporation rates stand at 0.13 inches and 0.2 inches respectively. Suppose the lake has a surface area of 1.87 acres and maintains a steady state, determine the annual average infiltration flow rate in ft³/hr.

(A) 0.6 ft³/hr
(B) 1.0 ft³/hr
(C) 1.6 ft³/hr
(D) 2.7 ft³/hr

9. In a lab assessment of BOD, 10 mL of wastewater was diluted to 200 mL. The initial DO concentration of the diluted sample measured was 10 mg/L, after 5 days of 20 °C incubation, the DO level was decreased to 4.6 mg/L. Assuming the wastewater has a BOD decay rate constant of 0.16 day^{-1} at base e, determine the ultimate BOD of the primary wastewater.

(A) 75 mg/L
(B) 108 mg/L
(C) 157 mg/L
(D) 196 mg/L

10. On St. Patrick's Day, Chicago has a tradition of dyeing the Chicago River green for celebration. The dye typically lasts a day or two and becomes invisible at a concentration of 1.00 mg/L. Assuming the river is a well-mixed batch reactor and the dye degrades at a rate constant of 10^{-2} per hour. Determine the initial dye concentration range: _____ - _____

(A) 1.04 mg/L
(B) 1.13 mg/L
(C) 1.27 mg/L
(D) 1.43 mg/L
(E) 1.62 mg/L
(F) 1.74 mg/L

11. A fire hydrant is designed to spray water at a flow rate of 1500 GPM during emergencies. If the maximum bearing pressure of the pitot gauge is 250 psi, what is the minimum square and sharp outlet diameter the fire hydrant needs?

(A) 1 in
(B) 2 in
(C) 4 in
(D) 6 in

12. A water treatment plant's water conveying network with flow rates labeled in GPM is shown below. Suppose streams A, B, and D have a pH of 8.0, 2.9, and 3.6 respectively, and reactors 1 and 2 are well-mixed. Determine the pH level of the effluent E.

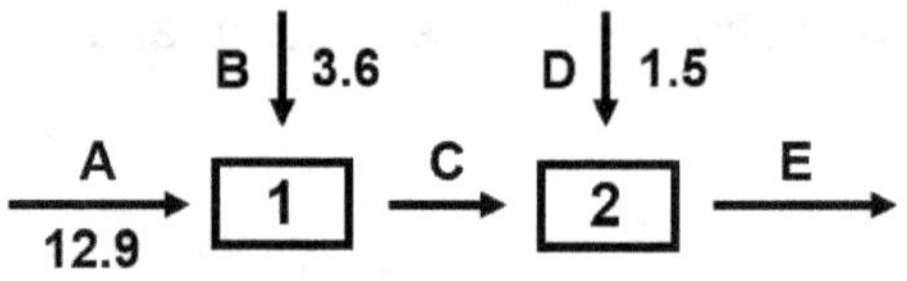

$$\xrightarrow[12.9]{A} \boxed{1} \xrightarrow{C} \boxed{2} \xrightarrow{E}$$

with B ↓ 3.6 entering reactor 1 and D ↓ 1.5 entering reactor 2.

(A) 3.6
(B) 4.2
(C) 6.6
(D) 6.9

13. A series pipe network is shown below. Assuming the system has a total head loss of 0.29 feet and both pipes feature a Darcy-Weisbach friction factor of 0.023, calculate the flow rate within the pipes.

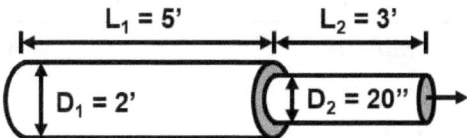

(A) 25 cfs
(B) 31 cfs
(C) 36 cfs
(D) 45 cfs

14. Water flows through a 4.8-inch diameter curved tube at a flow rate of 2.4 cfs under a pressure of 30 kPa as shown below. Neglect any potential head loss and the water self-weight, determine the magnitude of the reaction force exerted on the bend.

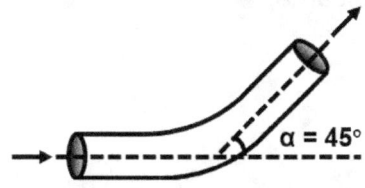

(A) 118.6 lbf
(B) 128.3 lbf
(C) 163.1 lbf
(D) 167.7 lbf

15. 100 mL wastewater was sampled to run the following experiment. Assume the wastewater has a generation rate of 0.8 MGD. Determine the solids loading of the system.

SAMPLE CONDITION	WEIGHT IN MILLIGRAM
EMPTY BEAKER	3121.5
SAMPLE IN BEAKER	93121.7
SAMPLE AFTER FULL WATER EVAPORATION	3294.1
SAMPLE AFTER 24HR 600°C FURNACE HEATED AND COOLED	3237.9

(A) 1.87 tons/day
(B) 3.88 tons/day
(C) 5.76 tons/day
(D) 7.76 tons/day

16. A 60-acre small urban features both business and residential catchment areas experience a rainfall event with an intensity of 2 inches per hour. Suppose the area ratio of business to residential is 3:7, and the runoff coefficients are 0.7 and 0.5 respectively, calculate the peak flow rate leaving the catchment area.

(A) 84 cfs
(B) 72 cfs
(C) 67 cfs
(D) 60 cfs

17. 200 lb sodium hypochlorous (molecular weight M = 74.44 g/mole) was added to a 100,000-gallon swimming pool for disinfection at a temperature of 25 °C. Determine the pH level of the pool water when the hydrolysis reaches the balance.

    Hint: The hypochlorous acid ionization constant $K_i = 3 \times 10^{-8}$ mole/L at 25 °C

    The water ionization constant $K_w = 1 \times 10^{-14}$ mole/L at 25 °C

    (A) 4.5
    (B) 4.6
    (C) 9.4
    (D) 9.5

18. A specific waterline is to be engineered for conveying groundwater with a tube at a steady flow rate of 7.3 cfs. Given the specifications of the concrete pipe provided below, calculate the minimum inlet pressure required to ensure the successful discharge of the liquid.

    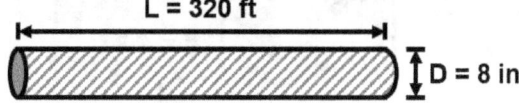

    (A) 19.8 psi
    (B) 22.7 psi
    (C) 24.5 psi
    (D) 28.1 psi

19. A carcinogenic substance is found in a county's drinking water system at a concentration of 4.2 mg/L. Suppose a 72 kg man has lived in the county for 70 years, drinking the contaminated water at a rate of 3.7 L/day, and the substance has a risk factor of 0.58 kg·day/mg. Determine the man's probability of risk from this exposure.

(A) 0.125
(B) 0.114
(C) 0.129
(D) 0.093

20. A diver submerged into a dead sea illustrated below and found the pressure at point A to be 3.65 psi. Determine the specific gravity of dead seawater based on the information provided.

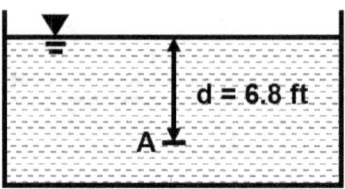

(A) 0.95
(B) 1.24
(C) 1.32
(D) 1.86

21. A water treatment plant removes TCE from a stream through activated carbon (AC) adsorption following the Langmuir Isotherm. Suppose that AC has a maximum adsorption capacity of 788 mg/g, and the Langmuir equilibrium constant is 2.4 L/µg. Determine the amount of AC needed to decrease the TCE concentration from 283 µg/L to 5 µg/L where the equilibrium happens.

(A) 0.38 mg/L
(B) 0.51 mg/L
(C) 0.69 mg/L
(D) 0.73 mg/L

22. A pumping well within a 38-foot thick confined aquifer is pumping at a constant rate of 33.6 gpm. The aquifer has a hydraulic conductivity of 12.8 ft/day and its original piezometric surface is 60 feet above the impermeable layer. Suppose drawdowns of 16.8 feet and 10.4 feet are observed in and 18 feet apart from the well. Determine the pumped well diameter.

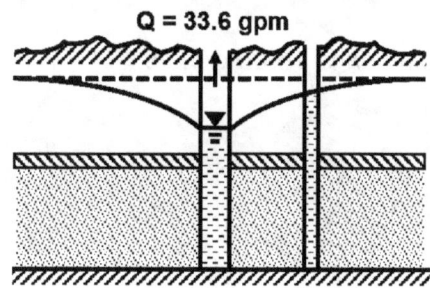

(A) 5.4 inches
(B) 10.6 inches
(C) 10.8 inches
(D) 21.2 inches

23. Point source contaminants from three local wastewater treatment plants are discharged into a lake as shown below. The effluent contaminant concentrations for sites X, Y, and Z are 7.8 mg/L, 3.6 mg/L, and 9.4 mg/L, respectively. Assuming the primary mechanism for contaminant spreading is diffusion, determine the direction of the chemical flux in the lake area.

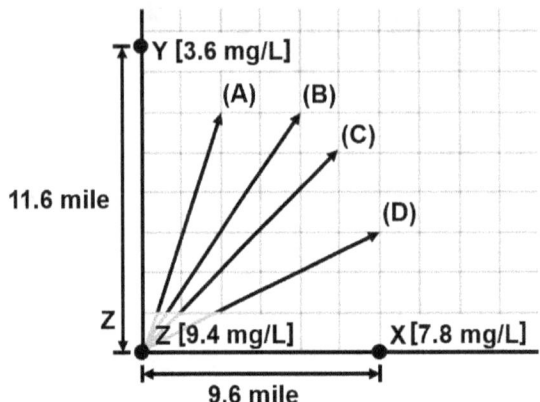

24. An industry discharges its effluent with a DO concentration of 2.43 mg/L to a nearby stream at a flow rate of 50 m³/s. The stream itself has a flow rate of 400 m³/s and is saturated with oxygen before the mixing occurs. Assuming the Henry's Law constant for oxygen is 0.0013 mole/L·atm and oxygen occupies 21% of the local air volume, calculate the oxygen deficit in the stream after mixing.

Hint: The atomic mass of oxygen is 16.00 g/mole.

(A) 0.22 mg/L
(B) 0.49 mg/L
(C) 0.70 mg/L
(D) 0.91 mg/L

25. Surface water in an open channel flows through a sharp-crested rectangular weir shown below. Given the dimensions provided, determine the discharge flow rate over the weir.

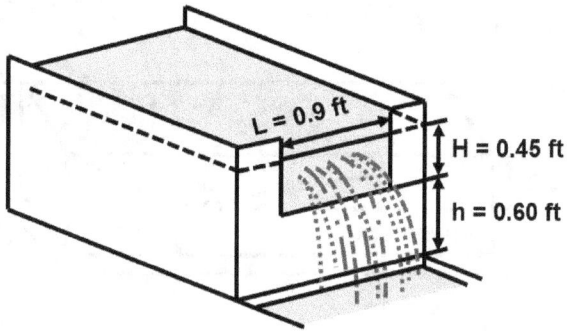

(A) 0.84 cfs
(B) 0.97 cfs
(C) 1.16 cfs
(D) 1.42 cfs

26. An air quality enhancement facility purifies polluted air by settling suspended particulate matter with a diameter of 50 μm and a density of 1640 kg/m³ in a settling chamber. Suppose the chamber has an airflow rate of 0.3 m³/s and a cross-sectional area of 3.7 m²; the air stream has an average density of 1.225 kg/m³ and a viscosity of 1.81×10⁻⁵ N·s/m². Determine the terminal settling velocity of the particles using Stoke's law.

(A) 0.03 m/s
(B) 0.08 m/s
(C) 0.12 m/s
(D) 0.15 m/s

27. The composite pavement and gutter system shown below is designed to facilitate runoff without disrupting traffic flow. Given an average Manning's roughness coefficient of 0.013, determine the flow rate of the channel.

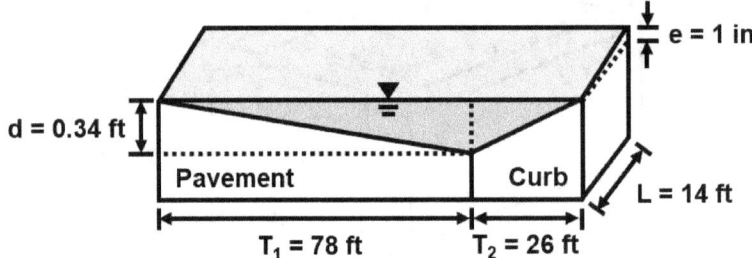

(A) 14 cfs
(B) 72 cfs
(C) 43 cfs
(D) 57 cfs

28. A tap water sample is analyzed to contain the following ions. Determine the total water hardness (mg/L as $CaCO_3$).

Hint: The atomic mass of chlorine is 35.5 g/mole.

IONS	CONCENTRATION
Ca^{2+}	16 mg/L
Na^+	36 mg/L
Mg^{2+}	?
Cl^-	84 mg/L
SO_4^{2-}	72 mg/L

(A) 34 mg/L
(B) 85 mg/L
(C) 115 mg/L
(D) 142 mg/L

29. A storm event lasted for 2.4 hours with a constant rainfall intensity of 3.5 millimeters per hour. The rainwater drains into the soil at an initial rate of 7.8 mm/hr with a decay factor of 1.45 per hour. Assuming the soil has a saturated conductivity of 2.8 mm/hr, determine the amount of rainwater that cannot be drained through soil infiltration at the end of the storm.

(A) 0 mm
(B) 1.7 mm
(C) 3.2 mm
(D) 8.4 mm

30. A horizontal pipe with changing diameters shown below is used to convey water from a reservoir to a drinking water treatment plant. The flow rate in the pipe is 1.8 m³/s, and the density of water to be treated is 1.03 g/m³. Determine the resultant force exerted on the pipe during the water transfer process.

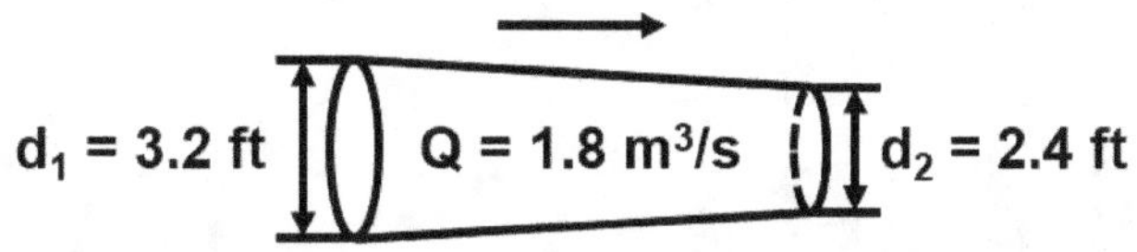

$d_1 = 3.2$ ft $Q = 1.8$ m³/s $d_2 = 2.4$ ft

(A) 109 lbf
(B) 781 lbf
(C) 1004 lbf
(D) 3491 lbf

31. Wastewater organic matter is to be treated with microorganisms in a 250 ft³ mixed batch reactor. Given the microorganisms have a maximum specific growth rate of 0.6 day^{-1}, a half-velocity constant of 30 mg/L, and a maintenance coefficient of 0.14 day^{-1}, determine the proper steady-state wastewater flow rate to obtain an effluent organic matter level of 20 mg/L.

(A) 0.29 gpm
(B) 0.31 gpm
(C) 0.25 gpm
(D) 0.13 gpm

32. A circular filtration system is installed for screening solid waste for a water treatment plant. Given a backwash flow rate of 1.24 ft³/s, a fluidized bed porosity of 0.47, and a terminal sedimentation velocity of 0.37 ft/s, determine the necessary diameter of the filter.

(A) 3.0 ft
(B) 4.4 ft
(C) 8.1 ft
(D) 11.5 ft

33. A coagulation-flocculation processing plant applies a turbine mixer with 6 curved blades to blend wastewater. The impeller diameter is 0.5 meters and rotates at 200 rounds per minute. Assuming the wastewater has a density of 1.16 g/cm³, calculate the power required for the impeller mixing process.

(A) 5.4 kW
(B) 6.5 kW
(C) 8.6 kW
(D) 9.3 kW

34. The fluctuation in rainfall intensity throughout a one-hour storm event is shown below. Assuming the local terrain characterizes a curve number of 87, calculate the runoff depth in inches using the NRCS method.

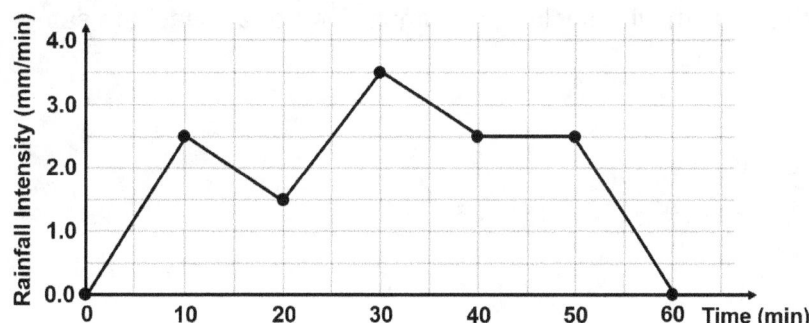

(A) 1.5 in
(B) 2.8 in
(C) 3.5 in
(D) 4.9 in

35. In a 1500-gallon activated sludge treatment reactor with a retention time of 6.3 hours, the suspended solids concentration is measured at 2100 ppm. Calculate the organic loading rate necessary to sustain a balanced system with a food-to-microorganism ratio of 0.32.

(A) 672 mg/L·day
(B) 256 mg/L·day
(C) 800 mg/L·day
(D) 531 mg/L·day

36. A technician is testing a 15 mL landfill leachate sample for odor control. After adding 35 mL of pure water to the raw sample, the odor still existed, so he transferred 15 mL of the diluted sample to another tube and added 24 mL of pure water until the odor was no longer perceptible. What is the threshold odor number for the raw leachate sample?

(A) 2.6
(B) 3.9
(C) 4.5
(D) 8.7

37. A small town has a current population of 16,000 and a saturation population of 25,000. Assuming the town has a growth rate constant of 5.6%, matching the time required for the population to reach its highest capacity with the population projection models provided.

38. Water flows from reservoir 1 to cylinder tubing 2 through a circular pipe as shown in the figure below. Assuming friction losses are negligible in the entire system, calculate the pipe flow velocity (v) with the following information provided:

Depth difference between sections: b
Depth of water in Reservoir 1: H
Depth of water in Tubing 2: h
Diameter of Tubing 2: D
Diameter of the Pipe: d
Water density: ρ

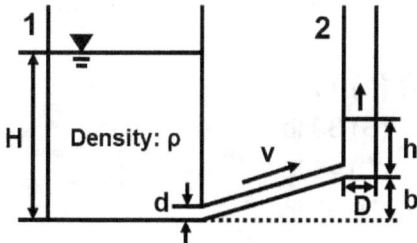

(A) $D^2/d^2 \cdot \sqrt{2g(H-h-b)}$
(B) $D^2/d^2 \cdot \sqrt{2\rho g(H-h-b)}$
(C) $D^2/d^2 \cdot \sqrt{2\rho g(H-h)}$
(D) $d^2/D^2 \cdot \rho\sqrt{2g(H-h)}$

39. A rainfall event produced 2.1 inches of precipitation in 25 minutes. Determine the probability of the event with the same or higher intensity happening only once within the next five years given the IDF curve shown below.

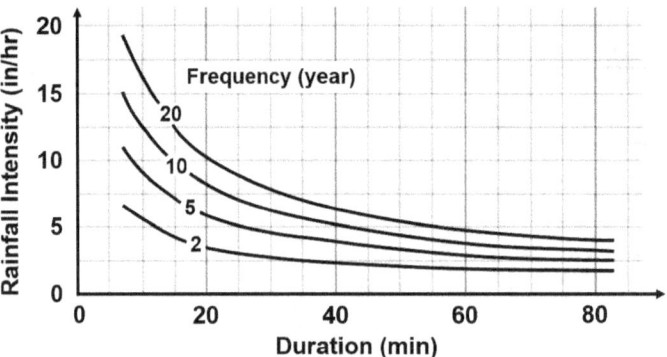

(A) 20%
(B) 41%
(C) 67%
(D) 80%

40. An industrial wastewater treatment plant applies an oxidation powder with an effective component of potassium permanganate ($KMnO_4$) to remove water impurities. According to calculations, the desired $KMnO_4$ dosage is 0.02 mmol/L. Assuming the powder has a purity of 90% and the plant has a flow rate of 1.63 MGD, determine the daily powder requirement.
Hint: The molecular weight of $KMnO_4$ is 158 g/mol.

(A) 11 lb
(B) 39 lb
(C) 48 lb
(D) 90 lb

Water Depth Solutions

Answer Keys

Question	Answer	Question	Answer
1.	B	21.	A
2.	C	22.	D
3.	C	23.	A
4.	A	24.	C
5.	D	25.	B
6.	C	26.	C
7.	B	27.	D
8.	B	28.	C
9.	D	29.	A
10.	C; E	30.	B
11.	B	31.	D
12.	A	32.	D
13.	C	33.	B
14.	B	34.	C
15.	C	35.	A
16.	C	36.	D
17.	D	37.	N/A
18.	B	38.	A
19.	A	39.	B
20.	B	40.	C

Solution 1

A triangular channel shown below has a flow rate of 125 ft³/sec and an average depth of 3 meters. What is the approximate critical depth of the channel?

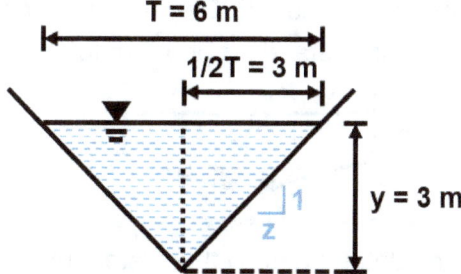

The critical depth (y_c) for triangle channels can be calculated through the formula:
$$y_c = (\sqrt{2}Z_c/z)^{0.4}$$

critical flow section factor $Z_c = Q/\sqrt{g} = 125 \div \sqrt{32.17} = 22.04$

angle ratio $z = (1/2T)/y = (1/2 \times 6m) \div 3m = 1$
critical depth $y_c = (\sqrt{2}Z_c/z)^{0.4} = (\sqrt{2} \times 22.04 \div 1)^{0.4} = 3.96$ft

The answer is **(B)**

Solution 2

In an area characterized by commercial and business areas situated on shallow loess sandy loam, a watershed with a length of 20 meters and a slope of 0.16% is utilized for stormwater hydrological management. Calculate the time of concentration for this watershed.

The time of concentration (t_c) can be determined through $t_c = 5/3 t_L$, where the time lag t_L:
$$t_L = 0.000526 L^{0.8} (\frac{1000}{CN} - 9)^{0.7} S^{-0.5}$$

The curve number (CN) for commercial and business areas featuring B-type hydrologic soil group (shallow loess, sandy loam) is 92 according to the reference manual: $CN = 92$;

watershed length $L = 20m = 65.6$ft
time lag t_L
$$t_L = 0.000526 L^{0.8} (\frac{1000}{CN} - 9)^{0.7} S^{-0.5} = 0.000526 \times 65.6^{0.8} (\frac{1000}{92} - 9)^{0.7} \times (0.16\%)^{-0.5} = 0.579 \text{hr}$$
time of concentration $t_c = 5/3 t_L = 5 \div 3 \times 0.579$hr $= 0.965$hr $= 57.9$min

The answer is **(C)**

Solution 3

A simplified city wastewater treatment facility for phosphorous (P) removal is shown below. Assuming the influent has a flow rate of 750.0 GPM with a P concentration of 6.3 mg/L, and the waste stream contains 184.6 mg/L of sediment P. Determine the effluent flow rate required when the P concentration meets the discharge standard of 0.1 mg/L.

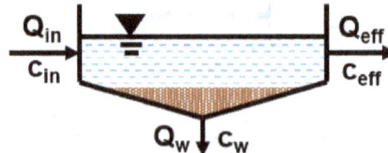

The effluent flow rate (Q_{out}) can be determined by the system mass balance relationships: 1. The phosphorous mass balance; 2. The flow rate balance. Two equations can be listed:

$$\begin{cases} Q_{in}c_{in} = Q_w c_w + Q_{eff} c_{eff} \\ Q_{in} = Q_w + Q_{eff} \end{cases}$$

Plug all the numbers in, we can have:

$$\begin{cases} 750 \text{gpm} \times 6.3 \text{mg/L} = Q_w \times 184.6 \text{mg/L} + Q_{eff} \times 0.1 \text{mg/L} \\ 750 \text{gpm} = Q_w + Q_{eff} \end{cases}$$

$$\begin{cases} Q_w = 25.2 \text{gpm} \\ Q_{eff} = 724.8 \text{gpm} \end{cases}$$

The answer is **(C)**

Solution 4

A triangular unit runoff hydrograph for a 160-acre watershed is shown below. Suppose the soil infiltration can be ignored, determine the peak discharge per centimeter of precipitation.

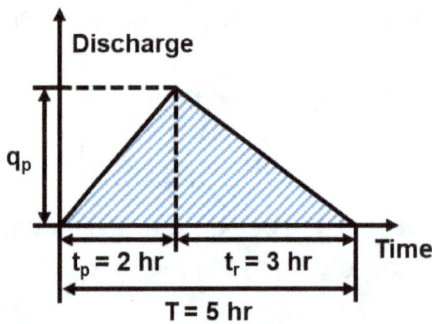

Peak discharge (q_p) can be calculated through the formula: $Q = 1/2 q_p (t_p + t_r)$
Considering one centimeter of effective precipitation, the total runoff volume Q (peak area)

$$Q = A \cdot d = 160 \text{acre} \times 1 \text{cm} = 160 \text{acre} \times 0.3937 \text{in} = 63.0 \text{acre} \cdot \text{in}$$

peak discharge q_p

$$q_p = \frac{2Q}{t_p + t_r} = \frac{2 \times 63.0 \text{acre} \cdot \text{in}}{2\text{hr} + 3\text{hr}} = 25.2 \text{acre} \cdot \text{in/hr}$$

Since the unit conversion from $\text{acre} \cdot \text{in/hr}$ to ft^3/s is most approximated to 1.0, $q_p = 25.2 \text{ft}^3/\text{s}$

The answer is **(A)**

Solution 5

A rectangular primary clarifier is 36 feet long and 24 feet 6 inches wide. The average influent flow rate to the clarifier is 490 GPM with a total suspended solids (TSS) concentration of 1.2 g/L. Considering the typical removal efficiency, what would be the effluent TSS concentration?

The effluent TSS concentration (C_{eff}) can be determined through the formula: $C_{eff} = C_{in} \cdot (1 - R)$
The typical primary clarifier removal efficiency (R) can be determined through the attached chart offered in the reference manual:

Typical Primary Clarifier Efficiency Removal

	Overflow Rates			
	1,200 (gpd/ft²) 48.9 (m/day)	1,000 (gpd/ft²) 40.7 (m/day)	800 (gpd/ft²) 32.6 (m/day)	600 (gpd/ft²) 24.4 (m/day)
Suspended Solids	54%	58%	64%	68%
BOD₅	30%	32%	34%	36%

In this question, the overflow rate $v_0 = Q/A$
surface area $A = L \cdot W = 36ft \times (24 + 6 \div 12)ft = 882ft^2$
flow rate $Q = 490gpm = 490gallon/min \times 60min/hr \times 24hr/day = 705600gpd$
overflow rate $v_0 = Q/A = 705600gpd \div 882ft^2 = 800gpd/ft^2$
According to the reference chart, when the overflow rate is $800gpd/ft^2$, the suspended solids removal efficiency R = 64%, therefore, the effluent TSS concentration C_{eff}
$C_{eff} = C_{in} \cdot (1 - R) = 1.2g/L \times (1 - 64\%) = 0.432g/L = 432mg/L$

The answer is **(D)**

Solution 6

Diesel oil is pumped from the storage tank to the ground gas station 10 feet above. Assuming the outlet oil pressure is 30 psi and the oil flow rate remains 8.3 GPM (nozzle diameter 0.86 inches), determine the pump break horsepower when the pump has an efficiency of 95%.

The pump break horsepower (BHP) can be determined through the formula:
$$BHP = \frac{Q \cdot H \cdot SG}{3956 \cdot \eta}$$

According to the reference manual, the specific gravity of diesel oil $SG = 0.85$;
specific weight of diesel oil $\gamma = \gamma_w \cdot SG = 62.4 \text{lbf/ft}^3 \times 0.85 = 53.04 \text{lbf/ft}^3$
total head added by the pump (H) can be calculated with the Bernoulli Equation:
$$H = z_a + \frac{P_a}{\gamma} + \frac{v_a^2}{2g}$$

nozzle area $A = \pi D^2/4 = 3.14 \times [0.86\text{in} \div (12\text{in/ft})]^2 \div 4 = 4.034 \times 10^{-3} \text{ft}^2$

velocity v_a
$$v_a = \frac{Q_a}{A} = \frac{8.3 \text{gallon}}{\text{min}} \times \frac{0.134 \text{ft}^3}{\text{gallon}} \times \frac{\text{min}}{60\text{s}} \div (4.034 \times 10^{-3} \text{ft}^2) = 4.6 \text{ft/s}$$

pressure $P_a = 30\text{psi} = 30 \text{lbf/in}^2 \times 144 \text{in}^2/\text{ft}^2 = 4320 \text{lbf/ft}^2$

total head added by the pump H
$$H = z_a + \frac{P_a}{\gamma} + \frac{v_a^2}{2g} = 10\text{ft} + \frac{4320 \text{lbf/ft}^2}{53.04 \text{lbf/ft}^3} + \frac{(4.6 \text{ft/s})^2}{2 \times 32.2 \text{ft/s}^2} = 91.8 \text{ft}$$

pump break horsepower BHP
$$BHP = \frac{Q \cdot H \cdot SG}{3956 \cdot \eta} = \frac{8.3 \text{gpm} \times 91.8 \text{ft} \times 0.85}{3956 \times 95\%} = 0.172 \text{hp}$$

The answer is **(C)**

Solution 7
Water is discharged from a sluice gate shown below. A hydraulic jump is observed to occur at Point A where the flow velocity is 18 ft/s at the depth of 2.4 meters. Determine the hydraulic jump classification from Point A to B.

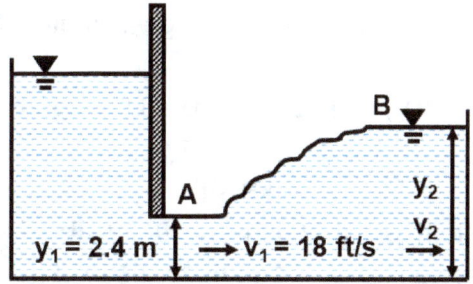

The classification of a hydraulic jump can be determined through the Froude number (Fr_1) at the upstream supercritical flow location, which can be calculated through the formula:

$$\frac{y_2}{y_1} = \frac{1}{2}(\sqrt{1 + 8Fr_1^2} - 1)$$

flow depth at Point A $y_1 = 2.4m \times 3.28ft/m = 7.87ft$

The flow depth at Point B (y_2) can be determined with the formula:

$$y_2 = -\frac{1}{2}y_1 + \sqrt{\frac{2v_1^2 y_1}{g} + \frac{y_1^2}{4}} = -\frac{1}{2} \times 7.87ft + \sqrt{\frac{2 \times (18ft/s)^2 \times 7.87ft}{32.17ft/s^2} + \frac{(7.87ft)^2}{4}} = 9.26ft$$

flow depth at Point B $y_2 = 9.26ft$

Inputting all the data into the Froude number (Fr_1) equation:

$$\frac{y_2}{y_1} = \frac{1}{2}\left(\sqrt{1 + 8Fr_1^2} - 1\right) = \frac{9.26ft}{7.87ft}$$

Froude number $Fr_1 = 1.13$

Since the Froude number (Fr_1) is within the range of $Fr_1 = 1.0 - 1.7$, the hydraulic jump should be classified as the Undular Jump.

The answer is **(B)**

Solution 8

In a given year, a natural lake experienced a surface inflow of 2,000 m³ and a drainage outflow of 1,600 m³. The average documented monthly precipitation and evaporation rates stand at 0.13 inches and 0.2 inches respectively. Suppose the lake has a surface area of 1.87 acres and maintains a steady state, determine the annual average infiltration flow rate in ft³/hr.

The average infiltration flow rate (Q_I) can be determined through the following formulas:
$$\begin{cases} P + Q_{in} - Q_{out} - E_s - I = \Delta S \\ I = Q_I t \end{cases}$$

surface water flow into the system $Q_{in} = 2000 m^3 = 70629.3 ft^3$
surface water flow out of the system $Q_{out} = 1600 m^3 = 56503.5 ft^3$
precipitation $P = d_p A = 0.13 in/month \times 12 month \times 1.87 acre = 10589.4 ft^3$
surface evaporation $E_s = d_s A = 0.2 in/month \times 12 month \times 1.87 acre = 16291.4 ft^3$
Since the lake maintains a steady state, the change in water storage $\Delta S = 0$;
Inputting all the data into the equations:
$$\begin{cases} 10589.4 ft^3 + 70629.3 ft^3 - 56503.5 ft^3 - 16291.4 ft^3 - I = 0 \\ I = Q_I \times (365 day \times 24 hr/day) \end{cases} \rightarrow \begin{cases} I = 8423.8 ft^3 \\ Q_I = 0.96 ft^3/hr \end{cases}$$

The answer is **(B)**

Solution 9

In a lab assessment of BOD, 10 mL of wastewater was diluted to 200 mL. The initial DO concentration of the diluted sample measured was 10 mg/L, after 5 days of 20 °C incubation, the DO level was decreased to 4.6 mg/L. Assuming the wastewater has a BOD decay rate constant of 0.16 day⁻¹ at base e, determine the ultimate BOD of the primary wastewater.

The ultimate BOD concentration (L_0) can be determined through the formula:
$$BOD_t = L_0(1 - e^{-kt})$$

incubating time $t = 5 day$
BOD testing concentration for 5 days BOD_5
$$BOD_5 = \frac{D_1 - D_2}{P} = \frac{10 mg/L - 4.6 mg/L}{10 mL/200 mL} = 108 mg/L$$

Inputting the data into the L_0 equation:
$$108 mg/L = L_0(1 - e^{-0.16 day^{-1} \times 5 day}) \rightarrow L_0 = 196 mg/L$$

The answer is **(D)**

Solution 10

On St. Patrick's Day, Chicago has a tradition of dyeing the Chicago River green for celebration. The dye typically lasts a day or two and becomes invisible at a concentration of 1.00 mg/L. Assuming the river is a well-mixed batch reactor and the dye degrades at a rate constant of 10^{-2} per hour. Determine the initial dye concentration range.

Since the dye degrades at a rate constant $k = 10^{-2} hr^{-1}$, as indicated by the unit of (hr^{-1}), the dye degradation is a first-order reaction. For first-order reactions, the following relationship exists:
$$\ln(C_A/C_{A0}) = -kt \quad \rightarrow \quad C_{A0} = C_A \cdot e^{kt}$$
If the dye lasts one day, time $t_1 = 24hr$, the initial dye concentration C_{A0-1d}
$$C_{A0-1d} = C_A \cdot e^{kt_1} = 1mg/L \times e^{10^{-2}hr^{-1} \times 24hr} = 1.27 mg/L$$
If the dye lasts two days, time $t_2 = 48hr$, the initial dye concentration C_{A0-2d}
$$C_{A0-2d} = C_A \cdot e^{kt_2} = 1mg/L \times e^{10^{-2}hr^{-1} \times 48hr} = 1.62 mg/L$$

Therefore, the initial dye concentration should be within the range of 1.27mg/L to 1.62mg/L, the answer is **(C)** and **(E)**

Solution 11

A fire hydrant is designed to spray water at a flow rate of 1500 GPM during emergencies. If the maximum bearing pressure of the pitot gauge is 250 psi, what is the minimum square and sharp outlet diameter the fire hydrant needs?

The outlet diameter (D) of the fire hydrant can be determined through the formula:
$$Q = 29.8 D^2 C_d P^{1/2} \quad \rightarrow \quad D = [Q/(29.8 C_d P^{1/2})]^{1/2}$$
Because the outlet is square and sharp, the hydrant coefficient $C_d = 0.80$, therefore we have:
$$D = [Q/(29.8 C_d P^{1/2})]^{1/2} = [1500 gpm \div (29.8 \times 0.80 \times (250 psi)^{1/2})]^{1/2} = 2.0 in$$
Since the pressure detected by the pitot gauge (p) should always be less than or equal to 250 psi:
$$p \leq P = 250 psi$$
The hydrant outlet diameter (d) should at least be 2.0 inches:
$$d \geq D = 2.0 in$$

The answer is **(B)**

Solution 12

A water treatment plant's water conveying network with flow rates labeled in GPM is shown below. Suppose streams A, B, and D have a pH of 8.0, 2.9, and 3.6 respectively, and reactors 1 and 2 are well-mixed. Determine the pH level of the effluent E.

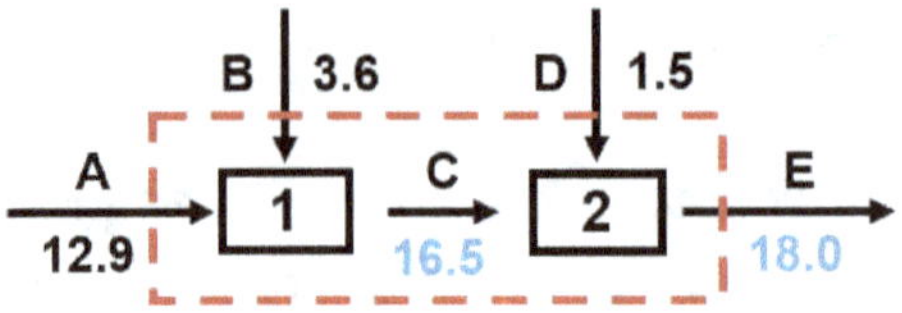

To calculate the pH level of the E, the flow rates on streams C and E need to be determined first:
Flow rate on stream C $Q_C = Q_A + Q_B = 12.9\text{gpm} + 3.6\text{gpm} = 16.5\text{gpm}$
Flow rate on stream E $Q_E = Q_C + Q_D = 16.5\text{gpm} + 1.5\text{gpm} = 18.0\text{gpm}$
The proton concentrations $[H^+]$ can be determined through the pH level on each stream:
stream A $[H^+]_A = 10^{-8.0}M$; stream B $[H^+]_B = 10^{-2.9}M$; stream D $[H^+]_D = 10^{-3.6}M$

According to the proton mass balance for the circled part, we have the following equation:
$$Q_A[H^+]_A + Q_B[H^+]_B + Q_D[H^+]_D = Q_E[H^+]_E$$
$12.9\text{gpm} \times 10^{-8.0}M + 3.6\text{gpm} \times 10^{-2.9}M + 1.5\text{gpm} \times 10^{-3.6}M = 18.0\text{gpm} \times [H^+]_E$
$[H^+]_E = 2.73 \times 10^{-4}M \quad \rightarrow \quad pH_E = -\log_{10}[H^+]_E = -\log_{10}(2.73 \times 10^{-4}) = 3.56$

The answer is **(A)**

Solution 13

A series pipe network is shown below. Assuming the system has a total head loss of 0.29 feet and both pipes feature a Darcy-Weisbach friction factor of 0.023, calculate the pipe flow rate.

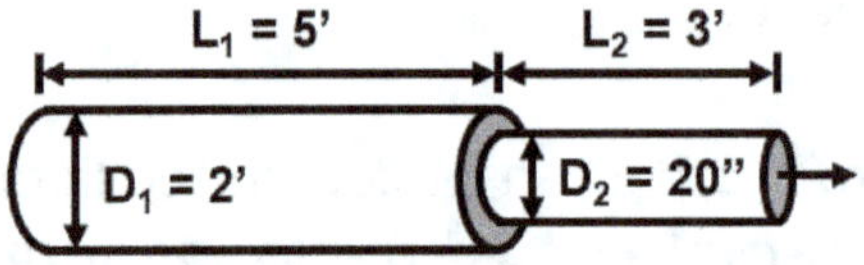

The head loss of the pipes can be calculated with the Darcy-Weisbach Equation: $h_f = fLv^2/2Dg$
In this case, the total head loss ($h_{f,tot}$) includes two different parts:
$$h_{f,tot} = h_{f,1} + h_{f,2} = f\frac{L_1}{D_1}\frac{v_1^2}{2g} + f\frac{L_2}{D_2}\frac{v_2^2}{2g}$$
Since the flow rate (Q) remains the same during the water-conveying process, we have:
$Q = v_1 \cdot \pi D_1^2/4 = v_2 \cdot \pi D_2^2/4 \quad \rightarrow \quad v_1 D_1^2 = v_2 D_2^2 \quad \rightarrow \quad v_1/v_2 = D_2^2/D_1^2 = 25/36$
Plug all the data into the head loss equation:
$$h_{f,tot} = f\frac{L_1}{D_1}\frac{v_1^2}{2g} + f\frac{L_2}{D_2}\frac{v_2^2}{2g} = 0.023 \cdot \frac{5\text{ft}}{2\text{ft}} \cdot \frac{(25/36 v_2)^2}{2 \times 32.2\text{ft/s}^2} + 0.023 \cdot \frac{3\text{ft}}{20/12\text{ft}} \cdot \frac{v_2^2}{2 \times 32.2\text{ft/s}^2} = 0.29\text{ft}$$
Solve the equation, we have $v_2 = 16.43\text{ft/s}$
flow rate $Q = v_2 \cdot \pi D_2^2/4 = 16.43\text{ft/s} \times 3.14 \times (20/12\text{ft})^2 \div 4 = 35.8\text{ft}^3/\text{s} = 35.8\text{cfs}$

The answer is **(C)**

Solution 14

Water flows through a 4.8-inch diameter curved tube at a flow rate of 2.4 cfs under a pressure of 30 kPa as shown below. Neglect any potential head loss and the water self-weight, determine the magnitude of the reaction force exerted on the bend.

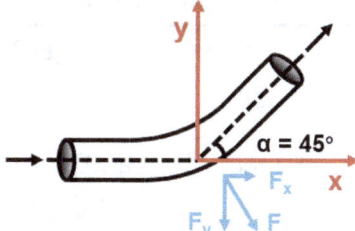

The reaction force exerted on the bend (F) can be determined in the x and y directions (F_x, F_y) respectively according to the reference manual:

$$\begin{cases} F_x = P_1A_1 - P_2A_2\cos\alpha - Q\rho(v_2\cos\alpha - v_1) \\ F_y = W + P_2A_2\sin\alpha + Q\rho(v_2\sin\alpha - 0) \end{cases}$$

Since the head loss and water self-weight are negligible, pipeline internal pressure $P_1 = P_2 = 30 \text{kPa} = 4.35\text{psi}$; the weight of the liquid $W = 0$;
cross-sectional area $A_1 = A_2 = \pi D^2/4 = 3.14 \times (4.8\text{in})^2 \div 4 = 18.1\text{in}^2 = 0.126\text{ft}^2$
water density $\rho = 62.4\text{pcf} = 1.94\text{slug/ft}^3$
velocity of the water flow $v_1 = v_2 = Q/A = 2.4\text{ft}^3/\text{s} \div 0.126\text{ft}^2 = 19.1\text{ft/s}$

Inputting all the data into the equations:

$$\begin{cases} F_x = 4.35\text{psi} \times 18.1\text{in}^2 \times (1 - \cos 45°) - 2.4\text{cfs} \times 1.94\text{slug/ft}^3 \times 19.1\text{ft/s} \times (\cos 45° - 1) \\ F_y = 0 + 4.35\text{psi} \times 18.1\text{in}^2 \times \sin 45° + 2.4\text{cfs} \times 1.94\text{slug/ft}^3 \times 19.1\text{ft/s} \times (\sin 45° - 0) \end{cases}$$

$$\begin{cases} F_x = 49.11\text{lbf} \\ F_y = 118.56\text{lbf} \end{cases}$$

combined reaction force F

$$F = \sqrt{F_x^2 + F_y^2} = \sqrt{(49.11\text{lbf})^2 + (118.56\text{lbf})^2} = 128.33\text{lbf}$$

The answer is **(B)**

Solution 15

100 mL wastewater was sampled to run the following experiment. Assume the wastewater has a generation rate of 0.8 MGD. Determine the solids loading of the system.

SAMPLE CONDITION	ACRONYME	WEIGHT IN MILLIGRAM
EMPTY BEAKER	Empty Beaker (EB)	3121.5
SAMPLE IN BEAKER	Sample Beaker (SB)	93121.7
SAMPLE AFTER FULL WATER EVAPORATION	Evaporate sample (ES)	3294.1
SAMPLE AFTER 24HR 600°C FURNACE HEATED AND COOLED	Furnace sample (FS)	3237.9

The solids loading rate (SL) can be calculated through the formula: $SL = 8.34QX$
suspended solids concentration X

$$X = \frac{W_{ES} - W_{EB}}{V_{sample}} = \frac{3294.1mg - 3121.5mg}{0.100L} = 1726 mg/L$$

solids loading rate $SL = 8.34QX = 8.34 \times 0.8 MGD \times 1726 mg/L = 11515.9 lb/day = 5.76 tons/day$

Please note that the "ton" unit used in the above equation refers to the US short ton.

The answer is **(C)**

Solution 16

A 60-acre small urban features both business and residential catchment areas experience a rainfall event with an intensity of 2 inches per hour. Suppose the area ratio of business to residential is 3:7, and the runoff coefficients are 0.7 and 0.5 respectively, calculate the peak flow rate leaving the catchment area.

The peak rainwater runoff discharge (Q) can be determined through the rational formula:
$$Q = C_w I A$$
weighted runoff coefficient C_w

$$C_w = \frac{A_1 C_1 + A_2 C_2}{A_1 + A_2} = (A_1\%)C_1 + (A_2\%)C_2 = 0.3 \times 0.7 + 0.7 \times 0.5 = 0.56$$

peak discharge $Q = C_w I A = 0.56 \times 2 in/hr \times 60 acre = 67.2 arce \cdot in/hr = 67.2 ft^3/sec$

The answer is **(C)**

Solution 17

200 lb sodium hypochlorous (molecular weight M = 74.44 g/mole) was added to a 100,000-gallon swimming pool for disinfection at a temperature of 25 °C. Determine the pH level of the pool water when the hydrolysis reaches the balance.

The hydrolysis reaction for sodium hypochlorous in water is:
$$ClO^- + H_2O \leftrightarrow HClO + OH^-$$
The equilibrium constant (K_H) of the above hydrolysis can be determined through the reactions:
$$HClO \leftrightarrow H^+ + ClO^- \quad K_i = 3 \times 10^{-8}$$
$$H_2O \leftrightarrow H^+ + OH^- \quad K_w = 10^{-14}$$
hydrolysis equilibrium constant $K_H = K_w/K_i = 10^{-14} \div (3 \times 10^{-8}) = 3.33 \times 10^{-7}$

sodium hypochlorous (NaClO) starting concentration c_s
$c_s = m/V = 200\text{lb} \div (10^5 \text{gal}) = 2 \times 10^{-3} \text{lb/gal} = 0.24\text{g/L} \div (74.44\text{g/mol}) = 3.22 \times 10^{-3} \text{mol/L}$
For the hydrolysis reaction:
$$ClO^- + H_2O \leftrightarrow HClO + OH^-$$
Starting: 3.22×10^{-3} 0   0
Balance: $3.22 \times 10^{-3} - x$ x   x
hydrolysis equilibrium constant K_H
$$K_H = \frac{[HClO][OH^-]}{[ClO^-]} = \frac{x \cdot x}{3.22 \times 10^{-3} - x} = 3.33 \times 10^{-7}$$
Solve the equation, $x = [OH^-] = 3.26 \times 10^{-5} M$

water pH level $pH = 14 + \log_{10}[OH^-] = 14 - 4.49 = 9.51$

The answer is **(D)**

Solution 18

A specific waterline is to be engineered for conveying groundwater with a tube at a steady flow rate of 7.3 cfs. Given the specifications of the concrete pipe provided below, calculate the minimum inlet pressure required to ensure the successful discharge of the liquid.

L = 320 ft

D = 8 in

The minimum inlet pressure (P) required to discharge the liquid equals the pressure head loss (P_l) during the pipe-conveying process, the pressure loss per unit length of pipe (P_f) can be calculated through the formula:

$$P_f = \frac{4.52 Q^{1.85}}{C^{1.85} D^{4.87}}$$

The flow rate $Q = 7.3 \text{cfs} = 3276 \text{gpm}$; the Hazen-Williams coefficient for concrete tubes $C = 130$
pressure loss per unit length of pipe P_f

$$P_f = \frac{4.52 Q^{1.85}}{C^{1.85} D^{4.87}} = \frac{4.52 \times 3276^{1.85}}{130^{1.85} \times 8^{4.87}} = 0.071 \text{psi/ft}$$

minimum inlet pressure P

$$P = P_l = P_f L = 0.071 \text{psi/ft} \times 320 \text{ft} = 22.72 \text{psi}$$

The answer is **(B)**

Solution 19

A carcinogenic substance is found in a county's drinking water system at a concentration of 4.2 mg/L. Suppose a 72 kg man has lived in the county for 70 years, drinking the contaminated water at a rate of 3.7 L/day, and the substance has a risk factor of 0.58 kg·day/mg. Determine the man's probability of risk from this exposure.

The probability of lifetime risk (p) can be determined through the formula:

$$p = CDI \times SF = C \frac{(IR)(EF)(ED)}{(BW)(AT)} SF$$

averaging time in days $AT = 70 \text{year} \times 365 \text{days/year} = 25550 \text{days}$
probability of lifetime risk p

$$p = C \frac{(IR)(EF)(ED)}{(BW)(AT)} SF = 4.2 \text{mg/L} \frac{3.7 \text{L/day} \times 365 \text{days/year} \times 70 \text{years}}{72 \text{kg} \times 25550 \text{days}} \times 0.58 \text{kg} \cdot \text{day/mg} = 0.125$$

The answer is **(A)**

Solution 20

A diver submerged into a dead sea illustrated below and found the pressure at point A to be 3.65 psi. Determine the specific gravity of dead seawater based on the information provided.

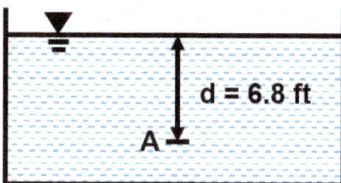

Since the pressure at Point A (P_A) can be determined through the formula $P_A = \rho g d$, the density of the dead seawater ρ

$$\rho = \frac{P_A}{gd} = \frac{3.65 \text{psi}}{9.8 \text{m/s}^2 \times 6.8 \text{ft}} = \frac{3.65 \times 6894.76 \text{pa}}{9.8 \text{m/s}^2 \times (6.8 \times 0.3048 \text{m})} = 1239 \text{kg/m}^3$$

the specific gravity of dead seawater SG

$$SG = \frac{\rho}{\rho_w} = \frac{1239 \text{kg/m}^3}{998 \text{kg/m}^3} = 1.24$$

The answer is **(B)**

Solution 21

A water treatment plant removes TCE from a stream through activated carbon (AC) adsorption following the Langmuir Isotherm. Suppose that AC has a maximum adsorption capacity of 788 mg/g, and the Langmuir equilibrium constant is 2.4 L/µg. Determine the amount of AC needed to decrease the TCE concentration from 283 µg/L to 5 µg/L where the equilibrium happens.

The amount of AC (C_{AC}) needed to remove the TCE can be determined through the formula:

$$q_e = \frac{C_0 - C_e}{C_{AC}} = \frac{283 \mu g/L - 5 \mu g/L}{C_{AC}} = \frac{278 \mu g/L}{C_{AC}} \rightarrow C_{AC} = \frac{278 \mu g/L}{q_e}$$

equilibrium loading on the activated carbon q_e

$$q_e = \frac{q_m K_L C_e}{1 + K_L C_e} = \frac{788 \text{mg/g} \times 2.4 \text{L/}\mu g \times 5 \mu g/L}{1 + 2.4 \text{L/}\mu g \times 5 \mu g/L} = 727.4 \text{mg/g}$$

amount of AC needed C_{AC}

$$C_{AC} = \frac{278 \mu g/L}{q_e} = \frac{278 \mu g/L}{727.4 \text{mg/g}} = 0.382 \text{mg/L}$$

The answer is **(A)**

Solution 22

A pumping well within a 38-foot thick confined aquifer is pumping at a constant rate of 33.6 gpm. The aquifer has a hydraulic conductivity of 12.8 ft/day and its original piezometric surface is 60 feet above the impermeable layer. Suppose drawdowns of 16.8 feet and 10.4 feet are observed in and 18 feet apart from the well. Determine the pumped well diameter.

Before calculating, we may illustrate the information provided to better understand the question:

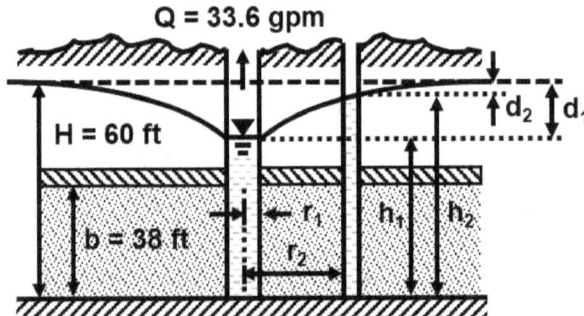

The radius of the well (r_1) can be determined through the following relationship first:

$$Q = \frac{2\pi Kb(h_2 - h_1)}{\ln(r_2/r_1)} \quad \rightarrow \quad r_1 = r_2 \cdot e^{-2\pi Kb(h_2-h_1)/Q}$$

The height difference of piezometric surfaces $\Delta h = h_2 - h_1$ can be transferred into the formula:

$$\Delta h = h_2 - h_1 = (H - d_2) - (H - d_1) = d_1 - d_2 = 16.8\text{ft} - 10.4\text{ft} = 6.4\text{ft}$$

hydraulic conductivity $K = 12.8\text{ft/day} = 1.48 \times 10^{-4}\text{ft/s}$
constant discharge rate $Q = 33.6\text{gpm} = 0.075\text{cfs}$
pumped well radius r_1

$$r_1 = r_2 \cdot e^{-2\pi Kb(h_2-h_1)/Q} = 18\text{ft} \cdot e^{-2\times3.14\times1.48\times10^{-4}\text{ft/s}\times38\text{ft}\times6.4\text{ft}\div0.075\text{cfs}} = 0.88\text{ft} = 10.6\text{in}$$

Therefore, the well diameter $d_1 = 2r_1 = 2 \times 10.6\text{in} = 21.2\text{in}$

The answer is **(D)**

Solution 23

Point source contaminants from three local wastewater treatment plants are discharged into a lake as shown below. The effluent contaminant concentrations for sites X, Y, and Z are 7.8 mg/L, 3.6 mg/L, and 9.4 mg/L, respectively. Assuming the primary mechanism for contaminant spreading is diffusion, determine the direction of the chemical flux in the lake area.

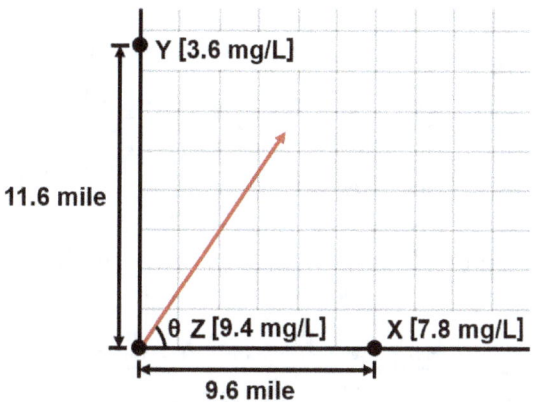

To determine the chemical flux direction, the dispersion mass flux in each direction (horizontal and vertical) should be calculated first

On the horizontal axis (along ZX):
dispersion mass flux $J_{disp-ZX} = D_h dc/d_x = D_h(9.4mg/L - 7.8mg/L) \div 9.6 mile = 0.167 D_h$

On the vertical axis (along ZY):
dispersion mass flux $J_{disp-ZY} = D_h dc/d_x = D_h(9.4mg/L - 3.6mg/L) \div 11.6 mile = 0.5 D_h$

where D_h is the hydrodynamic dispersion coefficient

Therefore, the inclination of the final flux to the horizontal axis $\tan\theta$

$$\tan\theta = \frac{J_{disp-ZY}}{J_{disp-ZX}} = \frac{0.5 D_h}{0.167 D_h} = 3$$

The answer is **(A)**

Solution 24

An industry discharges its effluent with a DO concentration of 2.43 mg/L to a nearby stream at a flow rate of 50 m³/s. The stream itself has a flow rate of 400 m³/s and is saturated with oxygen before the mixing occurs. Assuming the Henry's Law constant for oxygen is 0.0013 mole/L·atm and oxygen occupies 21% of the local air volume, calculate the oxygen deficit after mixing.

The oxygen deficit (D_a) can be determined through the formula:

$$D_a = DO_{stream} - \frac{Q_w DO_w + Q_{stream} DO_{stream}}{Q_w + Q_{stream}}$$

partial pressure of oxygen $P_{O_2} = P_{atm} \cdot \alpha_{O_2} = 1\,atm \times 0.21 = 0.21\,atm$
dissolved oxygen content in stream DO_{stream}
$DO_{stream} = K_H \times P_{O_2} = 0.0013\,mole/L \cdot atm \times 0.21\,atm \times 32\,g/mole = 8.74 \times 10^{-3}\,g/L = 8.74\,mg/L$
oxygen deficit D_a

$$D_a = DO_{stream} - \frac{Q_w DO_w + Q_{stream} DO_{stream}}{Q_w + Q_{stream}}$$

$$= 8.74\,mg/L - \frac{50\,m^3/s \times 2.43\,mg/L + 400\,m^3/s \times 8.74\,mg/L}{50\,m^3/s + 400\,m^3/s} = 0.70\,mg/L$$

The answer is **(C)**

Solution 25

Surface water in an open channel flows through a sharp-crested rectangular weir shown below. Given the dimensions provided, determine the discharge flow rate over the weir.

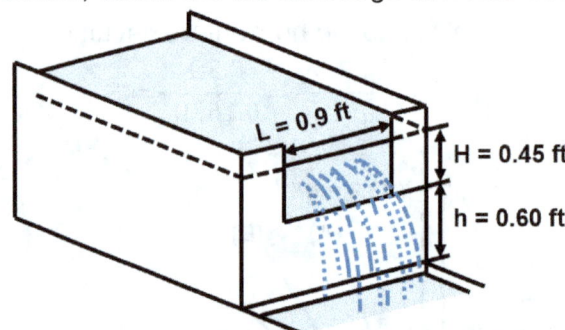

The discharge flow rate (Q) can be calculated through the formula: $Q = CLH^{3/2}$
The coefficient of discharge (C) in the above formula can be determined with the Rehbock Coefficient of Discharge Equation since the heights ratio $H/h = 0.45\,ft \div (0.60\,ft) = 0.75 < 10$

$$C = 3.27 + 0.4\frac{H}{h} = 3.27 + 0.4 \times \frac{0.45\,ft}{0.60\,ft} = 3.57$$

discharge flow over the wire $Q = CLH^{3/2} = 3.57 \times 0.9 \times (0.45)^{3/2}\,cfs = 0.97\,cfs$

The answer is **(B)**

Solution 26

An air quality enhancement facility purifies polluted air by settling suspended particulate matter with a diameter of 50 μm and a density of 1640 kg/m³ in a settling chamber. Suppose the chamber has an airflow rate of 0.3 m³/s and a cross-sectional area of 3.7 m²; the air stream has an average density of 1.225 kg/m³ and a viscosity of 1.81×10⁻⁵ N·s/m². Determine the terminal settling velocity of the particles using Stoke's law.

The terminal settling velocity (v_t) can be determined through Stoke's law formula:
$$v_t = \frac{g(\rho_p - \rho_g)d^2}{18\mu}$$

particle diameter $d = 50\mu m = 5 \times 10^{-5} m$
air absolute viscosity $\mu = 1.81 \times 10^{-5} N \cdot s/m^2 = 1.81 \times 10^{-5} kg/m \cdot s$
Please note the unit conversion $1N = 1kg \cdot m/s^2$ for the above equation.

terminal settling velocity v_t
$$v_t = \frac{g(\rho_p - \rho_w)d^2}{18\mu} = \frac{9.8 m/s^2 \times (1640 kg/m^3 - 1.225 kg/m^3) \times (5 \times 10^{-5} m)^2}{18 \times (1.81 \times 10^{-5} kg/m \cdot s)} = 0.123 m/s$$

The answer is **(C)**

Solution 27

The composite pavement and gutter system shown below is designed to facilitate runoff without disrupting traffic flow. Given an average Manning's roughness coefficient of 0.013, determine the flow rate of the channel.

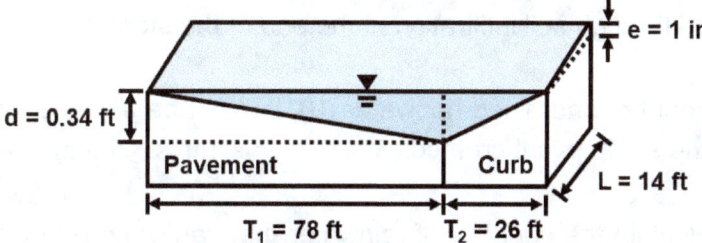

The flow rate (Q) of the channel can be determined through the equation:
$$Q = Q_1 + Q_2 = (0.56/n)(S^{0.5})(S_{x1}^{1.67} T_1^{2.67} + S_{x2}^{1.67} T_2^{2.67})$$
longitudinal slope $S = e/L = 1in \div (14ft \times 12in/ft) = 5.95 \times 10^{-3}$
pavement cross slope $S_{x1} = d/T_1 = 0.34ft \div (78ft) = 4.36 \times 10^{-3}$
curb cross slope $S_{x2} = d/T_2 = 0.34ft \div (26ft) = 0.013$
flow rate Q
$$Q = (0.56/n)(S^{0.5})(S_{x1}^{1.67} T_1^{2.67} + S_{x2}^{1.67} T_2^{2.67})$$
$$= (0.56 \div 0.013) \times [(5.95 \times 10^{-3})^{0.5}] \times [(4.36 \times 10^{-3})^{1.67} \times 78^{2.67} + 0.013^{1.67} \times 26^{2.67}] = 57 cfs$$

The answer is **(D)**

Solution 28

A tap water sample is analyzed to contain the following ions. Determine the total water hardness.

IONS	CONCENTRATION	CONCENTRATION
Ca^{2+}	16 mg/L	0.399 mM
Na^+	36 mg/L	1.565 mM
Mg^{2+}	?	?
Cl^-	84 mg/L	2.366 mM
SO_4^{2-}	72 mg/L	0.749 mM

To make the calculation easier, the concentrations for each ion are converted into molar units first; Since Ca^{2+} and Mg^{2+} ions are the primary causes of tap water hardness, the total hardness (TH) of the water sample can be calculated with the formula: $TH = [Ca^{2+}]_{CaCO_3} + [Mg^{2+}]_{CaCO_3}$
The Mg^{2+} concentration $[Mg^{2+}]$ can be determined through the charge conservation law:
$$2[Ca^{2+}] + [Na^+] + 2[Mg^{2+}] = [Cl^-] + 2[SO_4^{2-}]$$
$2 \times 0.399mM + 1.565mM + 2[Mg^{2+}] = 2.366mM + 2 \times 0.749mM \quad \rightarrow \quad [Mg^{2+}] = 0.751mM$
Therefore, the total hardness of the tap water TH
$TH = [Ca^{2+}]_{CaCO_3} + [Mg^{2+}]_{CaCO_3} = 0.399mM \times 100g/mol + 0.751mM \times 100g/mol = 115mg/L$

The answer is **(C)**

Solution 29

A storm event lasted for 2.4 hours with a constant rainfall intensity of 3.5 millimeters per hour. The rainwater drains into the soil at an initial rate of 7.8 mm/hr with a decay factor of 1.45 per hour. Assuming the soil has a saturated conductivity of 2.8 mm/hr, determine the amount of rainwater that cannot be drained through soil infiltration at the end of the storm.

To determine the amount of undrained rainwater (U), it might be easier to start with the amount of rainwater (S) that the storm event created: $S = t_p \cdot I = 2.4hr \times 3.5mm/hr = 8.4mm$

Then, we need to calculate the capacity of rainwater that can drain (F) during the storm event, which can be determined through the formula:
$$F = f_c t_p + \frac{f_0 - f_c}{k}(1 - e^{-kt_p}) = 2.8mm/hr \times 2.4hr + \frac{(7.8 - 2.8)mm/hr}{1.45/hr}(1 - e^{-1.45 \times 2.4}) = 10.1mm$$

There, we notice that the accumulated infiltration (F = 10.1mm) at the end of the storm is bigger than the amount of rainwater the storm created (S = 8.4mm), which means all the rainwater can infiltrated into the soil, leading to the amount of undrained rainwater U = 0

The answer is **(A)**

Solution 30

A horizontal pipe with changing diameters shown below is used to convey water from a reservoir to a drinking water treatment plant. The flow rate in the pipe is 1.8 m³/s, and the density of water to be treated is 1.03 g/m³. Determine the resultant force exerted on the pipe during the water transfer process.

$d_1 = 3.2$ ft $Q = 1.8$ m³/s $d_2 = 2.4$ ft

The resultant force exerted on the pipe (F_p) equals the resultant force acting on the fluid through the pipe (F_f) in magnitude, which equals the rate of change of momentum on the fluid (ΣF):

$$F_p = F_f = \Sigma F$$

Since we have $\Sigma F = F_{outlet} - F_{inlet} = \Sigma Q_2 \rho_2 v_2 - \Sigma Q_1 \rho_1 v_1$, the resultant force exerted on the pipe (F_p) can also be calculated through the formula: $F_p = F_{outlet} - F_{inlet}$

flow rate $Q = Q_1 = Q_2 = 1.8 m^3/s = 63.57 ft^3/s$
water density $\rho = \rho_1 = \rho_2 = 1.03 \times 62.4 pcf = 64.3 pcf$
inlet flow velocity $v_1 = Q/A_1 = 4Q/(\pi d_1^2) = 4 \times 63.57 ft^3/s \div [3.14 \times (3.2ft)^2] = 7.90 ft/s$
inlet rate of momentum F_{inlet}

$$F_{inlet} = \Sigma Q_1 \rho_1 v_1 = 63.57 ft^3/s \times 64.3 pcf \times 7.90 ft/s = 3.23 \times 10^4 lb \cdot ft/s^2 = 1004 lbf$$

outlet flow velocity $v_2 = Q/A_2 = 4Q/(\pi d_2^2) = 4 \times 63.57 ft^3/s \div [3.14 \times (2.4ft)^2] = 14.05 ft/s$
outlet rate of momentum F_{inlet}

$$F_{outlet} = \Sigma Q_2 \rho_2 v_2 = 63.57 ft^3/s \times 64.3 pcf \times 14.05 ft/s = 5.74 \times 10^4 lb \cdot ft/s^2 = 1785 lbf$$

pipe resultant force $F_p = F_{outlet} - F_{inlet} = 1785 lbf - 1004 lbf = 781 lbf$

The answer is **(B)**

Solution 31

Wastewater organic matter is to be treated with microorganisms in a 250 ft³ mixed batch reactor. Given the microorganisms have a maximum specific growth rate of 0.6 day⁻¹, a half-velocity constant of 30 mg/L, and a maintenance coefficient of 0.14 day⁻¹, determine the proper steady-state wastewater flow rate to obtain an effluent organic matter level of 20 mg/L.

The steady-state flow rate (Q) can be calculated through the formula: $Q = V/\tau$
The retention time (τ) can be determined with the Monod Kinetics:
$$\frac{1}{\tau} = \mu_{max}\frac{S}{K_s + S} - K_d = 0.6\text{day}^{-1} \times \frac{20\text{mg/L}}{30\text{mg/L} + 20\text{mg/L}} - 0.14\text{day}^{-1} = 0.1\text{day}^{-1} \rightarrow \tau = 10\text{day}$$
flow rate $Q = V/\tau = 250\text{ft}^3 \div (10\text{day}) = 25\text{ft}^3/\text{day} = 0.13\text{gpm}$

The answer is **(D)**

Solution 32

A circular filtration system is installed for screening solid waste for a water treatment plant. Given a backwash flow rate of 1.24 ft³/s, a fluidized bed porosity of 0.47, and a terminal sedimentation velocity of 0.37 ft/s, determine the necessary diameter of the filter.

The necessary filter diameter (D) can be determined through the following relationships shown in the reference manual:

$$n_f = \left(\frac{v_B}{v_t}\right)^{0.22}, \text{ where } v_B = \frac{Q_B}{A_{plan}} = \frac{Q_B}{1/4\pi D^2} = \frac{4Q_B}{\pi D^2}$$

$$n_f = \left(\frac{4Q_B}{\pi D^2 v_t}\right)^{0.22} \rightarrow D = \left(\frac{4Q_B n_f^{-1/0.22}}{\pi v_t}\right)^{0.5}$$

From the equation, we can have the filter diameter D

$$D = \left(\frac{4Q_B n_f^{-1/0.22}}{\pi v_t}\right)^{0.5} = \left(\frac{4 \times 1.24\text{ft}^3/\text{s} \times 0.47^{-1\div 0.22}}{3.14 \times 0.37\text{ft/s}}\right)^{0.5} = 11.5\text{ft}$$

The answer is **(D)**

Solution 33

A coagulation-flocculation processing plant applies a turbine mixer with 6 curved blades to blend wastewater. The impeller diameter is 0.5 meters and rotates at 200 rounds per minute. Assuming the wastewater has a density of 1.16 g/cm³, calculate the power required for the impeller mixing process.

To determine the power (P) required for the impeller mixer, the following formula can be applied:
$$P = K_T(n)^3(D_i)^5\rho_f$$
The impeller constant (K_T) can be determined with the "Values of the Impeller Constant Table" on the reference manual, given a turbine mixer with 6 curved blades in this question, $K_T = 4.80$
rotational speed $n = 200 \text{rev/min} = 3.33 \text{rev/sec}$
density of mixing solution $\rho_f = 1.16 \text{g/cm}^3 = 1.16 \times 10^3 \text{kg/m}^3$
impeller mixer power P
$P = K_T(n)^3(D_i)^5\rho_f = 4.80 \times (3.33 \text{rev/sec})^3 (0.5\text{m})^5 \times 1.16 \times 10^3 \text{kg/m}^3 = 6.44 \times 10^3 \text{w} = 6.44 \text{kw}$

The answer is **(B)**

Solution 34

The fluctuation in rainfall intensity throughout a one-hour storm event is shown below. Assuming the local terrain characterizes a curve number of 87, calculate the runoff depth in inches.

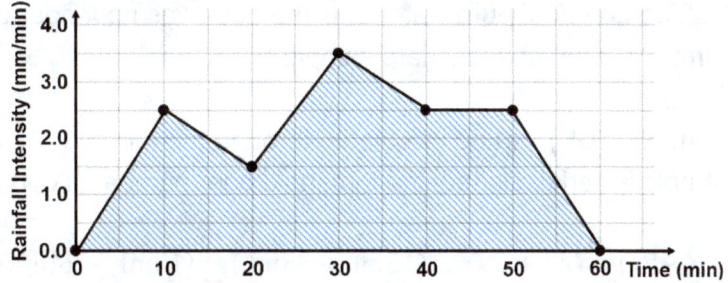

The runoff depth (Q) can be determined through the NRCS formula: $Q = (P - 0.2S)^2/(P + 0.8S)$
The precipitation depth (P) is the integral of rainfall intensity over time, which can be determined by calculating the area under the curve shown above: $P = [2.5 \times 10 \div 2 + (2.5 + 1.5) \times 10 \div 2 + (1.5 + 3.5) \times 10 \div 2 + (3.5 + 2.5) \times 10 \div 2 + 2.5 \times 10 + 2.5 \times 10 \div 2]\text{mm} = 125\text{mm} = 4.92\text{in}$
maximum basin retention $S = 1000/CN - 10 = 1000 \div 87 - 10 = 1.49\text{in}$
runoff depth Q

$$Q = \frac{(P - 0.2S)^2}{P + 0.8S} = \frac{(4.92\text{in} - 0.2 \times 1.49\text{in})^2}{4.92\text{in} + 0.8 \times 1.49\text{in}} = 3.5\text{in}$$

The answer is **(C)**

Solution 35

In a 1500-gallon activated sludge treatment reactor with a retention time of 6.3 hours, the suspended solids concentration is measured at 2100 ppm. Calculate the organic loading rate necessary to sustain a balanced system with a food-to-microorganism ratio of 0.32.

To maintain a certain food-to-microorganism ratio (F/M), the organic loading rate (L) relationship can be expressed through the following two formulas:

$$F/M = \frac{S_0 Q_0}{VX}, \quad L = \frac{Q_0 S_0}{V} \quad \rightarrow \quad L = (F/M) \cdot X$$

suspended solid concentration $X = 2100 \text{ ppm}$
organic loading rate $L = (F/M) \cdot X = 0.32 \text{ day}^{-1} \times 2100 \text{ ppm} = 672 \text{ mg/L} \cdot \text{day}$

The answer is **(A)**

Solution 36

A technician is testing a 15 mL landfill leachate sample for odor control. After adding 35 mL of pure water to the raw sample, the odor still existed, so he transferred 15 mL of the diluted sample to another tube and added 24 mL of pure water until the odor was no longer perceptible. What is the threshold odor number for the raw leachate sample?

The threshold odor number (TON) can be calculated through the formula: $TON = (A+B)/A$
Since the leachate sample is series diluted, the threshold odor number (TON) can be represented with another formula:

$$TON = \frac{(A_1 + B_1)}{A_1} \times \frac{(A_2 + B_2)}{A_2} = \frac{(15\text{mL} + 35\text{mL})}{15\text{mL}} \times \frac{(15\text{mL} + 24\text{mL})}{15\text{mL}} = 8.67$$

Alternatively, the threshold odor number (TON) can be determined through the original formula. The volume of odor-free water B contains not only $B_1 = 35\text{mL}$ for the initial dilution but also B_2' for the second time dilution. Since $B_2 = 24\text{mL}$ is for $A_2 = 15\text{mL}$ diluted sample, and we have $A_1 + B_1 = 15\text{mL} + 35\text{mL} = 50\text{mL}$ diluted sample in total, the corrected second dilution volume $B_2' = (50\text{mL}/15\text{mL}) \times 24\text{mL} = 80\text{mL}$; thus, the total water volume $B = 35\text{mL} + 80\text{mL} = 115\text{mL}$;

$$TON = \frac{A+B}{A} = \frac{15\text{mL} + 115\text{mL}}{15\text{mL}} = 8.67$$

The answer is **(D)**

Solution 37

A small town has a current population of 16,000 and a saturation population of 25,000. Assuming the town has a growth rate constant of 5.6%, matching the time required for the population to reach its highest capacity with the population projection models provided.

For linear growth, the required time (Δt_L) can be calculated through the formula: $P_t = P_0 + kP_0\Delta t_L$
elapsed time $\Delta t_L = (P_t - P_0)/(kP_0) = (25000 - 16000) \div (5.6\% \times 16000) = 10.04$ year
Please note that the linear projection formula provided in the reference manual is incorrect.

For exponential growth, the required time (Δt_E) can be calculated through: $P_t = P_0 e^{k\Delta t_E}$
elapsed time $\Delta t_E = \ln(P_t/P_0)/k = \ln(25000 \div 16000) \div (5.6\%) = 7.97$ year

For percent growth, the required time (Δt_P) can be calculated through: $P_t = P_0(1+k)^{\Delta t_P}$
elapsed time $\Delta t_P = \log_{1+k}(P_t/P_0) = \log_{1+5.6\%}(25000 \div 16000) = 8.19$ year

For decreasing rate growth, the required time (Δt_D) relationship: $P_t = P_0 + (S - P_0)(1 - e^{-k\Delta t_D})$
If $P_t = S$, the term $e^{-k\Delta t_D}$ should approach 0, therefore, elapsed time $\Delta t_D \to +\infty$

As a result, the answer is:

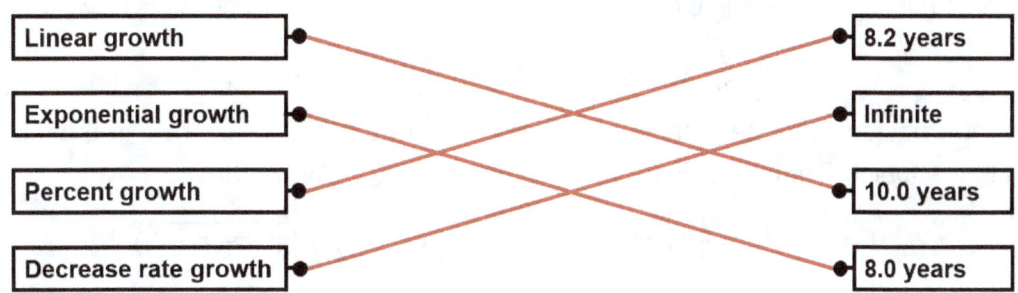

- Linear growth → 10.0 years
- Exponential growth → 8.0 years
- Percent growth → 8.2 years
- Decrease rate growth → Infinite

Solution 38

Water flows from reservoir 1 to cylinder tubing 2 through a circular pipe as shown in the figure below. Assuming friction losses are negligible in the entire system, calculate the pipe flow velocity (v) with the information provided:

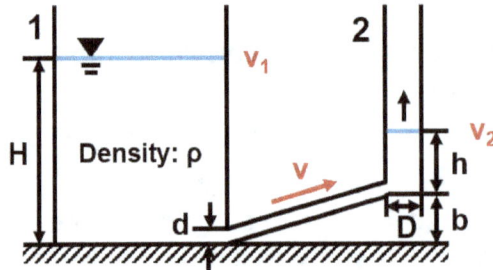

To determine the flow velocity (v) in the pipe, one way is to use the energy equation. Since the flow is one-dimensional and no pump or turbine exists between sections 1 and 2 in the system, the simplified field equation can be applied.

Taking the bottom of Reservoir 1 as the baseline, we have (from surface 1 to surface 2):

$$\frac{P_1}{\rho g} + \frac{v_1^2}{2g} + H = \frac{P_2}{\rho g} + \frac{v_2^2}{2g} + (h + b)$$

Since both sections are open to the air at the surface, the pressure on both sections (P_1, P_2) can be omitted as they both equal to the atmospheric pressure; also, the velocity of the water level drop at Reservoir 1 (v_1) is negligible considering its big surface area; there we have:

$$H = \frac{v_2^2}{2g} + (h + b) \quad \rightarrow \quad v_2 = \sqrt{2g(H - h - b)}$$

Because the transferring system shares the same flow rate (Q), the flow rate in the pipe should be equal to the tubing 2 flow rate. Based on the equation $Q = Av$, we have:

$$Q = \frac{1}{4}\pi d^2 v = \frac{1}{4}\pi D^2 v_2 \quad \rightarrow \quad v = \frac{D^2}{d^2} v_2 = \frac{D^2}{d^2} \sqrt{2g(H - h - b)}$$

The answer is **(A)**

Solution 39

A rainfall event produced 2.1 inches of precipitation in 25 minutes. Determine the probability of the event with the same or higher intensity happening only once within the next five years.

To determine the probability of a rainfall event that happens once in the next five years ($P_{1/5}$), it is important to figure out the probability of a single occurrence (p) first. To do that, we need to calculate the rainfall event intensity I = Precip/time = 2.1in ÷ [25min ÷ 60(min/hr)] = 5.04in/hr
Plotting the rainfall intensity (I = 5.04in/hr) and duration (D = 25min) into the IDF curve:

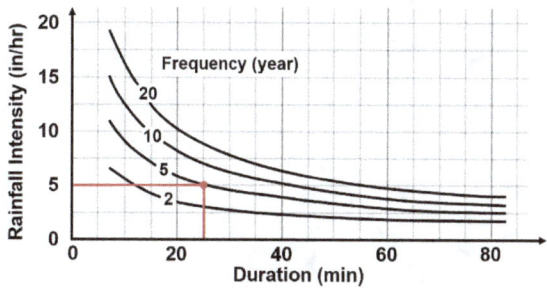

We can find that the frequency/return period (T) for the specific event T = 5yr, the probability of a single occurrence p = 1/T = 1 ÷ 5 = 0.2; Therefore, the probability of the rainfall event that happens once in the next five years $P_{1/5} = C_5^1 \times 0.2^1 \times (1 - 0.2)^4 = 0.4096 = 41\%$

The answer is **(B)**

Solution 40

An industrial wastewater treatment plant applies an oxidation powder with an effective component of potassium permanganate (KMnO₄) to remove water impurities. According to calculations, the desired KMnO₄ dosage is 0.02 mmol/L. Assuming the powder has a purity of 90% and the plant has a flow rate of 1.63 MGD, determine the daily powder requirement.

The daily powder requirement (m) can be calculated with the formula: m = cQt/η
potassium permanganate concentration c = 0.02mmol/L × 158g/mol = 3.16mg/L
flow rate Q = 1.63 × 10⁶ gallon/day = 6.17 × 10⁶ L/day
daily powder requirement m

$$m = \frac{cQt}{\eta} = \frac{3.16 \text{mg/L} \times 6.17 \times 10^6 \text{L/day} \times 1\text{day}}{90\%} = 2.17 \times 10^7 \text{mg} = 47.8\text{lb}$$

The answer is **(C)**

www.ingramcontent.com/pod-product-compliance
Lightning Source LLC
Chambersburg PA
CBHW082340220526

5470CB00008B/2576